D1325324

BLACKPOOL AND THE FYLDE COLLEGE
3 8049 00135 036 5

DEMOLITION: PRACTICES, TECHNOLOGY, AND MANAGEMENT

DEMOLITION: PRACTICES, TECHNOLOGY, AND MANAGEMENT

by
Richard J. Diven and Mark Shaurette

Purdue University Press
West Lafayette, Indiana

Printed in the United States of America.

ISBN 978-1-55753-567-2

Library of Congress Cataloging-in-Publication Data

Diven, Richard J., 1939-
Demolition : practices, technology, and management / by Richard J. Diven & Mark Shaurette.
p. cm.
Includes bibliographical references and index.
ISBN 978-1-55753-567-2
1. Wrecking. I. Shaurette, Mark, 1954- II. Title.
TH447.D58 2010
690'.26--dc22
2010014884

TABLE OF CONTENTS

FOREWORD

Demolition is as old as mankind. Since ancient times, man has recycled his cities and towns, reusing materials, revitalizing his environment, saving what is important, preparing for the "new."

The art and science of the demolition process can seem mysterious. Part magic, part engineering, it can be specific for every structure and locale. It can be labor-intensive or require state-of-the-art equipment specifically designed for demolition. It can involve delicate handwork or the crush of a wrecking ball. The equipment used by the modern demolition contractor is extremely sophisticated and very productive.

Demolition is multifaceted. It involves structural dismantlement, industrial recovery, salvage, recycling and reuse, implosion, specialized rigging, hazardous material handling, the management of toxic substances, disaster remediation, facilities decontamination, landfilling, project management, and general contracting.

Demolition requires knowledge of personnel, equipment, the nature of structures and architectural design, finance, real estate, recycling and salvage, occupational health and safety, environmental protection, site remediation, and project management. It is characterized by entrepreneurial spirits who love the process for its challenges.

The National Demolition Association (NDA) was founded in 1973 by forward-thinking demolition contractors interested in bringing professionalism to their industry. It has been in the forefront of representing the interests of everyone involved in the demolition process from the smallest demolition company to the Fortune 500 manufacturer of products and services for the industry.

This demolition textbook is the first attempt the NDA has ever made to produce a compendium of demolition means and methods in one volume. It is designed to educate future demolition contractors, estimators, project managers, superintendents, and foremen of the practical procedures used by the modern demolition entity. It also will provide the industry's client base of architects, engineers, scientists, developers, general contractors, and the like with a clear understanding of just how demolition is conducted in the twenty-first century.

The National Demolition Association has partnered with Purdue University's Department of Building Construction Management to meet the goals outlined in its mission of "providing the tools necessary to be leaders in environmental stewardship, safety, education, professional competency and government advocacy." We are grateful to the authors of this textbook, Richard J. Diven and Mark Shaurette, as well as the many contributors from the demolition industry for their hard work and diligence to meet these goals.

We anticipate that as the demolition process evolves there will be many additions to this work. It is our greatest hope that this book establishes a clearer understanding of the demolition process and enhances the professionalism of its practitioners.

Michael R. Taylor, CAE
Executive Director
National Demolition Association
February 1, 2010

PREFACE

The purpose of this textbook is to introduce engineers and managers to the basics of the demolition industry and to provide a reference guide to assist the inexperienced reader in understanding various facets of this very interesting business. Society is undergoing changes that will influence the future of demolition such as rebuilding our infrastructure and replacing buildings and structures originally built for technologies that have become obsolete.

As existing structures within the United States and Canada age, it is anticipated that opportunities in demolition and reconstruction will continue to expand. In a 2007 survey of owners responsible for facility construction and maintenance, the Construction Management Association of America (CMAA) outlined a set of seven challenges they believe will cause construction markets to change direction in the near future. The first challenge outlined indicated that "Aging infrastructure in nearly every market segment is at or beyond its current useful life...representing trillions of dollars in necessary spending over the next ten to twenty years to upgrade and replace these assets" (D'Agostino et al., 2007).

It is the authors' opinion that few construction related fields offer as much opportunity for innovative thinking and career development as does the demolition industry.

ACKNOWLEDGMENTS

The development and production of this textbook is the result of the contributions of a number of individuals who share the authors' enthusiasm for the fascinating field of demolition.

This textbook would not be possible were it not for the funding and leadership that was provided by the National Demolition Association (NDA), its Education Committee, and the NDA Executive Director, Michael R. Taylor, CAE.

Bill Moore, NDA past president, contributed significantly, providing industry advice on content as well as numerous photographs used throughout the text.

Other individuals who made significant contributions to content and editing include National Demolition Association members Jerry Myrick and Robert Klotzbach and current NDA president Raymond Passeno, CIH. Also providing valuable content information were Carrie Taylor, Francis Stohosky, and Michael Allen.

Photographs were provided by individuals and organizations too numerous to list. The enhancements provided by their contribution of visual material are greatly appreciated.

Many thanks for the cooperation of the Purdue University Department of Building Construction Management and its department head, Robert F. Cox, PhD.

PUBLISHER NOTE

Estimating spreadsheets and other electronic materials useful to readers of this book can be found in a companion digital collection: http://docs.lib.purdue.edu/demolition/

PHOTO CONTRIBUTION ACKNOWLEDGMENT

Kevin Behling
Peter Bigwood, Atlas Copco
Nicole Bock, DARDA GmbH
Jerry Curry
Richard Diven
Dan Hoffman, Allied-Gator, Inc.
Wanda LeBleu, EPA Office of Resource Conservation and Recovery
Stacey Loizeaux, Controlled Demolition, Inc.
Bill Moore, Brandenburg Industrial Service Co.
Randy Rapp
Mike Ramun, Allied Erecting & Dismantling Co., Inc.
Jim Redyke, Dykon Explosive Demolition Corp.
Tom Robinette, Robinette Demolition, Inc.
Mark Shaurette
P. Alex Shaurette
Jeff Slotta, J. Harper Contractors, Inc.

CHAPTER 1

INTRODUCTION TO THE INDUSTRY

A. IT'S NOT CONSTRUCTION

This book is designed to be a guide to demolition practice, technology, and management as it is practiced in North America. In 2008 there were approximately eight hundred companies in the United States and Canada whose main activities were classified as demolition. In addition to this group, there are many general contractors and specialty contractors that perform some demolition, usually in conjunction with their non-demolition projects. Even though demolition activities are often grouped into the general classification of construction, *it is not construction* and, in fact, is much different from most construction activities.

Differences include such tasks as performing the work in a more or less reverse order from construction and the requirement to ensure that hazardous materials have been surveyed and are removed. Tasks associated with removal of hazardous materials are seldom a concern in construction projects. Also, delivery and storage of materials is seldom a concern to the demolition contractor. The cost effect of a potentially large credit for salvage materials from a demolition job can be a significant component of the final net cost and is a determining factor in selecting the method of demolishing a building.

B. WHAT IS DEMOLITION?

In our modern world, demolition has evolved into a rather complex combination of tasks: from clearing sites for new construction to salvaging building materials from structures that have been demolished to removing hazardous materials. Therefore, we ask, "what kind of work does a demolition contractor do?" The following Table 1.01 is a list of typical tasks performed by demolition contractors and a brief explanation of these tasks.

Table 1.01

Demolition Type	Description
Commercial and Industrial Buildings and Structures—Low Rise Demolition	Any building/structure less than eight stories high; usually demolished using heavy equipment such as cranes and excavators equipped with special attachments.
Commercial and Industrial Buildings and Structures—High Rise	Any building over eight stories in height; often demolished using explosives if surrounding conditions and local regulations permit. For non-explosive demolition, floors above eighty feet are usually demolished one floor at a time, and lower floors are demolished with heavy equipment.
High Rise industrial structures such as chimneys, towers, chemical plants, steel mills, and similar structures	This category of demolition includes many special use buildings and structures. They are usually demolished using cranes, excavators, explosives, and various combinations of dismantling techniques.
Sub-grade, Concrete Foundation Demolition	Often the most expensive part of a demolition project is the breaking and removal of heavy, below grade concrete foundations. Large excavators equipped with hydraulic breakers are commonly used to break the concrete. The crane and wrecking ball procedure is also used for breaking foundations. Explosives are used when conditions are suitable.

Utility Demolition	This category can include any type of utilities either above or below grade and is usually a part of the primary demolition task.
Demolition of Parking Lots, Roads, and Runways	These improvements can be either concrete or asphalt and are commonly performed as part of structure demolition, but may also be stand-alone projects.
Bridge Demolition	Bridge demolition has become a major part of the demolition industry as our old bridges become obsolete. This category includes all types of bridges and materials used in their construction.
Railroad Demolition	Includes primarily the removal of abandoned tracks, appurtenant structures, and railroad bridges.
Interior Demolition	This is also a major category for demolition contractors and is typically the first step required for the processes of remodeling or upgrading existing buildings.
Selective Demolition	This term can be applied to most any of the above categories in the sense that "selective demolition" may be defined as removal of specified parts of buildings, structures, utilities, and process equipment. For example, the end wall of a hospital building may need to be removed to permit the construction of an addition to the building.
Explosive Demolition	Explosives used in demolition are typically used to fell buildings and structures or to break up heavy concrete foundations. The term "implosion" is used to describe the process of falling buildings in a controlled manner. A discussion of the use of explosives is included in chapter 11.
Marine Demolition	This category covers demolition of pilings, docks, piers and foundations, sunken vessels, and underwater obstructions.
Disaster Response	This is a broad category and includes cleanup of debris from hurricanes, earthquakes, fires, and floods. Included is the demolition of damaged structures and infrastructure. Also included are emergency rescues.
Historical Salvage	This includes the careful removal and salvage of items of historical importance and is often performed by specialists in this kind of work.
Hazardous Materials Removal	While not a direct demolition activity, the removal of hazardous materials such as asbestos-containing materials (ACM), polychlorinated biphenyls (PCBs), and other hazardous materials are performed as an integral phase of most demolition projects. Either the demolition contractor or a subcontractor will perform this work before demolition work begins. There are a number of regulations and laws governing such work.

C. BRIEF HISTORY OF DEMOLITION

When the ruins of Homer's ancient Troy were discovered by the German archeologist Heinrich Schliemann, it was determined that Troy was the seventh city to have been constructed on that site. English medieval church building materials often originated in earlier buildings that may have been deliberately demolished to supply materials. Many of these reused materials were recycled from previous structures of Roman origin. Although building stone was the most typically recycled material, timber, tiles, lead, and even plaster or mortar were recycled (Blair and Ramsay 1991). The systematic harvesting of building materials in medieval England was not unique. Hundreds of years earlier, Constantine destroyed city walls and monumental public buildings as punishment during his conquests, only to rebuild with reused building materials to reduce costs (Bowman, Garnsey, and Cameron 2005).

The ravages of time and conquest were not the only reason for demolition. In 1910, the Gillender Building in lower Manhattan, the tallest office structure in the world when constructed in 1897, was razed. This high-class structure just over twelve years old was demolished to make way for a more elaborate and prestigious structure. While the Gillender Building was demolished using over two hundred men over a period of forty-five days with little more than pneumatic tools and power winches, machine wrecking began in earnest in the 1930s when the wrecking ball, as we know it, became common. Early crane wrecking was an adaptation of the ancient demolition practice of ramming. The availability of cranes allowed demolition contractors to suspend heavy lengths of flat or square steel from a crane's cables and use them as a ram. Ball shaped rams of steel weighing thousands of pounds gradually replaced flat or square iron because these rams could be swung rather than simply dropped (Byles 2005). Cranes equipped with clamshell buckets and compressed air jackhammers soon joined the demolition contractor's arsenal of machinery to assist in razing structures.

Through much of the early twentieth century, power shovels were employed to load demolition debris when wooden chutes could not be used to convey debris directly into waiting trucks. Advances in the internal combustion engine and later high-output diesel engines allowed machine demolition to rapidly expand. After World War II, the pace of road building and the development of high-strength steel led to significant advancements in earth moving equipment that was well suited to use in the demolition industry. Although the wrecking ball and crane are still used today, advances in hydraulically powered equipment has to a large extent allowed the crane and wrecking ball to be replaced by track mounted excavators. These excavators employ a variety of hydraulically powered attachments to wreck buildings as well as to separate and load demolished materials for reuse, recycling, or landfill disposal.

The use of explosives in demolition is a dramatic process originally used for breaking structures into smaller more easily managed pieces. In the late 1950s, a more controlled use of explosives, often called "implosion," was first used. The term "implosion" indicates that the material collapses inward rather than flying outward as would be expected from an "explosion." When implosions are utilized, gravity plays a major role in the building's destruction. Small quantities of explosives placed in carefully planned positions throughout the structure are used to direct the building's collapse in a predictable direction. Carefully planned timing of the explosive charges assists in providing an orderly collapse and prevents damage from percussion that would occur if all charges were detonated at the same time. Although the spectacular nature of implosions make them well known to the general public, it has been estimated by the National Demolition Association that less than 1% of all modern demolition work in the United States utilizes explosives (National Demolition Association 2007).

During the 1800s, demolition contractors would often pay for the privilege of wrecking a building in return for the right to salvage the building materials. In cities, such as New York where new building activity was close and prevalent, bricks, plumbing fixtures, pipes, steel beams, marble, and granite were often salvaged. Reclaimed timber was frequently used for shoring or to construct a "canopy" around the building to protect the public. Over time, salvage became less profitable. The shift from lime- to Portland cement-based mortars made masonry structures more costly to demolish and made used brick or stone harder to clean. In addition, the rising cost of labor made reuse of most materials more costly to the point that, by the late 1920s, the economics of demolition salvage had reversed. By that time, demolition contractors were charging substantial sums to wreck old buildings. As an example, when the Hotel Majestic in the Central Park West area of New York was demolished in 1929, less than 10% of the wrecking expense was covered by salvage (Byles 2005).

In 1963, New York City's old Pennsylvania Station was demolished to make way for a new sports complex to be called Madison Square Garden. When Pennsylvania Station was constructed in 1910, Saint Michael's Church was painstakingly dismantled and rebuilt on another site to make way for the new train station. By contrast, when Pennsylvania Station was demolished, much of the ornate travertine statuary and four foot diameter granite columns from the station became swamp fill in the nearby New Jersey Meadowlands. Much has changed in demolition reuse and recycling since the 1960s. Architectural salvage is now a major business (Byles 2005). In addition, modern machinery

Figure C.01. One of the eagles from the 1910 Pennsylvania Station mounted near the new Pennsylvania Station. Only fourteen of the original twenty-two were salvaged during demolition.

has allowed a reduced dependence on labor. This reduced labor burden has again reversed the economics of salvage for many commodities. The profit potential resulting from careful salvage and the regulatory restriction on landfills in many areas has greatly reduced the proportion of demolition debris sent to landfills. Since the early 1990s, recycling has increased to become a major part of the demolition materials handling process. As a result, the demolition industry has become a leader in the improvement of our environment.

During the last two decades of the twentieth century, the demolition industry experienced substantial growth. During the 1990s, the number of demolition contractors grew nearly 60% with the dollar value of both payroll and the value of work performed increasing twofold. At the same time the industry became aware of the dangers of some of the materials produced during the demolition process and has worked with both OSHA and EPA to produce regulations designed to protect the health and safety of its employees and the public (Byles 2005).

D. STUDY QUESTIONS

1. List three examples of differences between demolition and general construction.

2. What was the major reason that structures in ancient civilization were demolished?

3. What type of demolition category would the following projects most resemble?
   a. Demolition of a fifteen story office building?
   b. Demolition of a railroad trestle?
   c. Demolition of bridge abutments?

4. Could the demolition contractor legally remove asbestos from a building that he or she had already wrecked without regulatory agency approval?

5. If the interior walls and ceilings were to be removed for remodeling, what are the two types of demolition that could be involved?

REFERENCES

D'Agostino, B., Mikulis, M., and Bridgers, M. 2007. *FMI & CMAA Eighth Annual Survey of Owners*. Raleigh, NC: FMI.

Blair, J., and Ramsay, N. 1991. *English Medieval Industries: Craftsmen, Techniques, Products*. London: Continuum International Publishing Group.

Bowman, A., Garnsey, P., and Cameron, A. 2005. *The Cambridge Ancient History: The Crisis of Empire, A. D. 193-337*. Cambridge: Cambridge University Press.

Byles, J. 2005. *Rubble: Unearthing the History of Demolition*. New York: Crown Publishing Group.

National Demolition Association. 2007. *10 Common Misconceptions about the Demolition Industry*.

CHAPTER 2

THE DEMOLITION CONTRACTOR

A. AS PRIME CONTRACTOR

A "prime contractor" may be defined "as the contractor who has directly contracted with the owner." The prime contractor may also be referred to as the "general contractor" if the prime has other contractors working under his authority and direction. The duties of the prime contractor include, but are not limited to, the sample duties listed below:

- Enter into a contract with the owner or his representative.
- Provide specified insurance and bonds for the owner's protection.
- Provide invoices for the owner.
- Contract with subcontractors as required.
- Retain responsibility for total oversight on the project, which includes managing the site safety plan, work schedule, and quality control.
- Negotiate with the owner for any changes in the contract tasks, pricing, and schedules.
- Receive payments from the owner and distribute payments to subcontractors and suppliers.

There are some projects that may be under the overall control of a non-demolition prime contractor but have a large enough demolition component that the demolition subcontractor operates as a "prime subcontractor." Such a contractual relationship is common on large federal projects such as the demolition and environmental cleanups of superfund sites. An example of such a project is the cleanup of the Hanford Nuclear Site in Washington state where the entire site and all projects were under the control of a large general contractor. For major demolition tasks associated with this project, a prime demolition subcontractor would fulfill all necessary tasks to complete the demolition requirements.

From a business standpoint, demolition contractors often prefer to be a prime contractor rather than a subcontractor. As a prime contractor, the demolition contractor is the first to get paid and has control over all site operations defined in the contract specifications.

B. AS SUBCONTRACTOR

The demolition contractor will typically be a subcontractor on a larger project that includes the construction of new facilities. The demolition company, as a subcontractor, will operate under the terms of the contract between the owner and general contractor, and as such, the demolition contractor should demand a copy of any of the terms of the prime contact that could impact his work. In general, being a subcontractor has a number of undesirable qualities. When working as a subcontractor, the demolition contractor should keep the following in mind:

- Especially important is an understanding of the terms of payment and the scheduling of the work. As a subcontractor, the demolition company is no longer paid by the owner, but must wait until the general contractor is paid—sometimes this can be a real problem when payments are not made in a timely manner. It is also important for the demolition subcontractor to thoroughly understand the schedule of work and be certain that other subcontractors are not scheduled in such a manner as to interfere with his work and/or create safety hazards. When preparing a bid for a job, the subcontractor should clearly state any exceptions he or she may have to the specifications. Such exceptions may include clarifications such as terms of payment and including retention of funds (i.e., funds held back until the entire job is completed—usually 10%).

- Define any apparent schedule conflicts.
- Clarify who is responsible for removal of hazardous materials if not clearly set forth in the specifications.
- Clarify who has ownership of salvaged materials from the demolition.

There are positive reasons for a demolition contractor to seek a subcontractor role on some projects:

- A project, while having a large demolition component, may be too complex for the experience level of the demolition contractor.
- A demolition contractor may have had excellent experience with a general contractor and would rather have the general take care of all the administrative functions of the job including dealing with the owner.

C. WORK AREA—LOCAL—REGIONAL—NATIONWIDE

C.01. Local Work Area

Many smaller firms choose to limit their operations to the city in which they live or perhaps their state. Most of these firms are family-owned businesses that operate on a small scale compared to larger contractors. There are advantages to doing business in a local area insofar as the contractor will have a thorough knowledge of local regulatory personnel, permit costs, landfill costs, recycling opportunities, and other special conditions. Another advantage is the ability to maintain a relatively small overhead because fewer jobs are in progress at any one time.

The primary disadvantage is the periodic problems created by local economic slowdowns. The result is very competitive pricing, which can produce low or non-existent profits. If work does get slow in their area, the contractor may have to lay off good workers. This cycle makes it more difficult to start up a new job and obtain the experienced workers required. In addition, employees are less certain about their job security.

C.02. Regional Work Areas

The regional demolition contractor will frequently have more than one office location, especially when their region includes two or more states. The primary advantages for a contractor working in an extended geographic area are:

- More opportunities to bid jobs.
- Ability to keep working when another specific area encounters a slowdown in work.
- Temporarily moving employees to out-of-town projects allows the contractor to keep experienced people productive.
- Employees can have more secure careers.
- By moving equipment to where it is needed, the equipment will be best utilized.
- The regional contractor can better keep overhead costs under control by spreading management costs over more dollar volume.
- Access to broader recycling markets.

The primary disadvantage of working as a regional contractor is the problem of maintaining relatively large overhead and equipment costs when the work slows down in an entire region or nationally. Compared with the local contractor, there is an increased difficulty in maintaining a core of experienced personnel during a slowdown in the region.

C.03. Nationwide Work Area

Few demolition contractors have the ability and financing to operate as national firms. There are, however, several national firms maintaining regional offices with a large number of personnel and fleets of major equipment. All of the advantages and disadvantages of the regional firms listed above are magnified with a national firm. A major advantage for a contractor to work nationwide is the ability to bid and perform the largest and most complicated projects in several areas at one time. In addition, such firms have the resources and personnel to quickly respond to large disasters.

D. DEMOLITION PERSONNEL AND THEIR DUTIES

As with all companies, personnel are the demolition contractor's most valuable asset. In the demolition industry, it is extremely important that the contractor have people with the experience and talent in key positions. Since there are a wide variety of firms doing demolition work, not every company will require the same categories of personnel. The following list shows typical employment categories for a demolition company and a brief summary of the duties of each:

Project Estimators: These individuals estimate the cost of completing the project by reviewing the project's scope of work, analyzing detailed demolition/architectural/structural plans, and conducting a thorough visual inspection of the project. The estimator is responsible for interpreting the scope of work and specifications to determine the cost associated with each detail including permits and bonds, regulatory issues, manpower and equipment needs, and subcontractors. There are many factors that go into estimating a demolition project, and it is far more challenging than estimating most construction projects. A demolition project estimator will have an extensive knowledge of construction materials and means as well as methods of demolition in order to develop and estimate the cost of the demolition plan.

Project Superintendent or Supervisor: These individuals run the projects and make the day-to-day decisions of what is to be done and how it is to be done. They have overall responsibility for safety, quality control, operations, subcontractor interfaces, schedule compliance, and assignments of people and equipment as well as dealing with the owner as necessary. The superintendent must have sufficient experience in the type of demolition being performed and usually is required to be on the job at all times. On smaller projects, the superintendent may also be assigned the duties of safety officer and/or quality control officer.

Project Manager (PM): This individual is usually the primary contact between the owner and the project superintendent. Their duties include assisting the supervisor in his or her duties and managing the contract, negotiating changes in the contract, processing the periodic reports to the owner's representative, preparing documentation for invoicing payments, and on some jobs, may serve dual roles as either the site safety officer and/or quality control officer. Often one PM will have responsibility for more than one project, especially if those projects are not large.

Site Safety Officer (SSO): This individual monitors the safety compliance for one or more projects. Often this individual will be a corporate employee and have more than one project or region to service at any given time. For large or complicated projects, a safety officer is commonly assigned full-time responsibilities to one job. The site safety officer should have demolition-specific experience and specialized safety training in compliance with contract specifications and government regulations. A minimum of thirty hours of OSHA Safety Training will be mandatory for this person in addition to a recommended minimum of demolition experience.

Quality Control Officer (QCO): The QCO is seldom required for a typical demolition project unless there is substantial amount of new construction as a part of the contract. For demolition work, quality control and site safety are often performed by the same individual

Foremen: The demolition foremen are the key to a safe, smooth running, and efficient job. The foremen report directly to the superintendent and usually direct the activities of two to ten workers. Depending on the job, the crew sizes may be larger. Typically, a foreman will conduct regular meetings with his crew to inform them of what is to be done and how they are to do it. Often, a foreman is a working foreman, especially for smaller sized crews. Foremen are usually promoted from within a company and are well trained in the jobs assigned to their crews.

Equipment Operators: This labor category, also known as Operating Engineers, is the personnel that operate and maintain the heavy equipment. These individuals usually work under the supervision of a foreman. Successful demolition firms have learned the importance of providing steady employment for their operators. It can take years for an equipment operator to learn the skills required for today's complex demolition machines such as excavators, cranes, and loaders.

Laborers: Laborers are usually divided into two groups, skilled and unskilled. The skilled laborers perform critical jobs such as rigging loads to be lifted, cutting heavy steel beams with oxy-fuel (oxy-acetylene or propane) torches, operating pneumatic tools, and similar tasks. Unskilled laborers perform such tasks as using water hoses to control dust, sweeping streets, moving materials and small equipment, provide flagging services for traffic control, and a host of other tasks that do not require special skills.

Hazardous Materials Workers: This category of workers has both skilled and unskilled workers, and their tasks usually require special training. The subject of worker training is discussed in more detail in chapter 8.

Note: A discussion of typical administrative personnel is presented in chapter 7.

E. UNION AND NON-UNION LABOR RELATIONS

The decision for a demolition contractor to become either a union or a non-union company is a choice made by company management. Some contractors elect to have both union and non-union entities under their ownership.

The advantages of being a union contractor are summarized as follows:

- In most metropolitan areas, there is a pool of experienced workers available through a union hall.
- Many unions maintain schools for training their members.
- The union labor member has fringe benefits, including health insurance.
- The contractor may quickly add personnel to his workforce from the union hall.
- Some projects are restricted to union contractors because of agreements made by the general contractor and/or owners.

The disadvantages of being a union contractor are summarized as follows:

- The cost of labor is usually more expensive than non-union.
- The contractor is bound by work rules that may sometimes be quite restrictive.
- The union, not the contractor, can decide what wages and fringe benefits are paid to each employee.
- A labor union decision to go on strike can seriously cripple a contractor's activities.
- Union contractors may not work alongside non-union contractors.

F. STUDY QUESTIONS

1. List one or more reasons for a demolition contractor to prefer to work in a local area rather than a region.

2. List one or more reason that a demolition company may to prefer to work as a subcontractor rather than as a prime contractor.

3. As the owner of a medium sized demolition company, what key personnel would you assign to work on a complex industrial project?

4. What key personnel would you assign to satisfy requirements for demolishing a few wood frame houses?

CHAPTER 3

MODERN DEMOLITION PRACTICES

A. INTRODUCTION

The methods used for demolition vary greatly depending on many factors. The more important factors include the following considerations:

- Type of construction to be demolished
- Size of structure(s)
- Equipment available either by ownership or available as rentals
- Schedule requirements
- Limitations imposed by the specifications
- Availability of key personnel
- Workload of contractor

It is important to note that the type of demolition normally done in the U.S. and Canada may vary significantly to that done in Europe, Japan, and most other countries. Reasons for these differences include the following:

- More or less stringent regulatory requirements
- Availability of "state-of-the-art" equipment
- Cost of manual labor
- More or less stringent regulations for removing and handling asbestos materials and other hazardous materials
- Values and marketability of used materials from demolition projects
- Differences in recycling and disposal requirements of demolition materials

This chapter will consider only those demolition practices commonly used in North America.

B. ANALYZING THE DEMOLITION PROJECT

Before one can determine the approach to a particular demolition project, a number of factors must be considered. For the purpose of this sample analysis, we will assume that the project is of significant size and includes several different types of construction in an industrial setting. The contractor or his designated competent person such as the project superintendent or project manager must ask the following questions during the analytical process:

- How complex is the project?
- Is the project suitable for the company's operations?
- What submittals are required?
- How is safety and risk management to be controlled?
- How is the schedule to be developed, and who controls it?
- What equipment will be ideal for the project?
- How are materials, such as imported backfill, to be managed?

- What equipment is available from the contractor's own fleet; what needs to be rented?
- What key personnel are available?
- How will utility work be handled?
- How will subcontractor interfaces be handled?
- How will interfaces with other contractors working on the site be handled?
- Who is responsible for hazardous materials removal?
- What preparations need to be made for waste management and to remove recyclable materials?

Depending upon the specifications for the project, the analytical process can be more or less extensive.

C. PREPARATORY TASKS

Prior to commencing work, the demolition contractor must ensure that the appropriate preparatory tasks have been assigned to experienced personnel and that the schedule for completing such tasks is accomplished. Improper preparation can result in a number of serious problems after the project has begun. For example, failure to properly address the safety issues for a particular project can result in serious exposure to injury or property damage during the course of the work. Or, failure to have a proper permit could result in a job being delayed and significant costs being incurred. The following sample checklist of preparatory tasks is important to setting up and maintaining control over the various aspects of the project:

C.01. Administrative

- Have submittals required by specifications been submitted and approved?

  __Y __N

- Has the OSHA Engineering Survey required by 29 CFR 1926 Part T, 850(a) been submitted?

  __Y __N

- Have subcontracts been negotiated and signed?

  __Y __N

- Have labor unions been contacted, if appropriate?

  __Y __N

- Are all permits in order?

  __Y__N

- Are reports and billing instructions assigned to appropriate personnel?

  __Y __N

C.02. Operations

- Have all necessary arrangements been made to remove asbestos and other hazardous materials—either with own employees or a subcontractor?

  __Y__N

- Have utilities been located and arrangements made for capping or re-routing?

  __Y __N

- Have all arrangements for moving heavy equipment to the jobsite been accomplished?

  __Y __N

- Have all items identified in the OSHA Engineering Survey been addressed and plans developed for correcting any problems identified in the Survey (e.g., providing an engineered shoring plan for building elements in danger of collapse)?

  __Y __N

- Have provisions for supplying fuel been made?

  __Y __N

- If a jobsite office is needed, have arrangements been made?

  __Y __N

- Have potential buyers of scrap metals and other salvage been contacted?

  __Y __N

The chances of conducting a safe and profitable job are greatly increased if close attention is paid to the initial preparation and planning. It has been said that the "devil is in the details," and this is so true for a demolition project.

D. MACHINE DEMOLITION

By far, most of today's demolition operations rely on heavy equipment to get the job done. Advances in construction machinery engineering have been responsible for the development of a wide variety of specialized demolition equipment. Perhaps the most significant development has been the evolution of high pressure hydraulics to transfer the machine's power to the attachment tools such as shears, breakers, concrete pulverizers, and others. These machines and their attachments are discussed in more detail in chapter 9.

Progressive wrecking of buildings and structures three stories or less is commonly done with either excavators or front-end loaders. The machine operator can work at a safe distance and either chew the building apart or pull or push it over to be processed by segregating the materials and loading. Except for the use of explosives and the felling of a building, most forms of demolition may be described as a progression of repetitive wrecking techniques. Figure D.01 is an example of a typical excavator, with bucket and thumb, using the method of "progressive wrecking" to demolish an industrial building.

Figures D.01 and D.02 are both examples of the progressive wrecking of low rise buildings.

The *progressive demolition of high-rise buildings and structures* is accomplished using one or a combination of the three principal methods listed below:

(1) Machines with high-reach booms can be effective up to over one hundred feet in height (Figure D.03). Modern high-reach demolition machines have become a standard method for the demolition of buildings up to about twelve stories. This method has a proven safety record and can be accomplished in areas of restricted access. The larger excavators of fifty to seventy-five tons are fitted with "third member" booms to allow them to safely work at heights of over 120 feet or about thirty-seven meters. Attachments such as shears, grapples, and concrete "crackers" are used, and the operator is usually in a tilting cab so that he or she can easily see the work. Video cameras can be mounted on the boom tips to provide the operator with a view of what he or she is doing. This method is of wrecking is relatively quiet, and using water spray nozzles directed from the boom to the work area can significantly reduce dust problems.

(2) The crane and suspended wrecking ball have been the standard for demolition of buildings and structures of twelve stories or less since the early 1950s (Figure D.04). The wrecking ball can be dropped or swung to strike the structure, and the operator is far enough away that he or she is

Figure D.01. Pulling over wall in progressive demolition.

Figure D.02. Wheel loader with grapple bucket.

Figure D.03. High-reach demolition tool with grapple.

protected from falling demolition debris. If the crane is duty cycle equipped with a third cable drum, the cable from that drum can be attached to the wrecking ball, pulled back, and then released to accomplish a very controlled method of demolition. The wrecking balls are made of cast steel and weigh anywhere from 1,500 to 10,000 pounds. (Note: Swinging the wrecking ball over any improvements not designated for demolition is prohibited is some states.)

(3) For buildings over twelve stories and for buildings that do not have sufficient operating room for safe demolition by excavators, cranes, or the use of explosives, the standard method is demolition of one floor at a time using a variety of small skid steer loaders and excavators (Figure D.05). The "floor-by-floor" method for demolishing buildings was the first system used for bringing down buildings without using explosives until the crane and ball or crane and clamshell bucket came into widespread use. (Note: Use of the clamshell bucket is rarely used in modern demolition—this is discussed in chapter 9). The basic system is to place small equipment such as skid steer loaders and mini-excavators on the top floor of a building using either cranes outside the building or hoists set up inside the building to lift the equipment to the working level. Depending upon the type of construction and the structural integrity of the building, the small machines break up concrete and masonry, and steel is cut with torches. The building is literally dismantled or deconstructed piece-by-piece. Debris is usually dropped to lower levels through elevator shafts where it can be loaded into trucks. A crane can also be used to lower skids (steel boxes) filled with debris.

Large structures can be either "tripped" or "pulled" under the following conditions: (1) sufficient space is available to avoid damage to other structures; (2) the fall zone will be of sufficient distance from any underground constructions, such as sewers, to avoid impact damage; (3) the building to be felled is structurally sound enough to eliminate the possibility of an unplanned collapse during preparation.

Tripping a structure is a means to bring down the entire structure. (Note: Explosives are also a means of bringing down high buildings and structures, and their use is discussed in chapter 11.) Tripping is accomplished by weakening the structure's columns and shear walls and bracing in a systematic way so that workers are not exposed to the danger of an unplanned collapse. Columns are weakened by pre-cuts using handheld torches or a shear. A crane, using a wrecking ball, can knock out the weakened columns, thus allowing the building to fall in a predetermined direction.

Pulling a building or structure is another method of bring down an entire structure at once. This is typically accomplished by attaching heavy cables to columns at high points on the structure to achieve a significant mechanical advantage, then removing shear walls, bracing at the ground level, and weakening columns similar to that required for tripping a structure. When all is ready, the pre-attached cables are connected to one or more pieces of heavy machinery located well beyond the "fall zone." The cables are pulled in the direction of fall with the heavy equipment. As soon as the center of gravity is moved beyond the outer wall line, gravity takes over.

Foundations and below-grade structures such as basements or vaults are demolished in a variety of ways, but again the excavator is the principal machine. It can be used both for the excava-

Figure D.04. Crawler crane with wrecking ball.

Figure D.05. Floor-by-floor demolition.

tion and to break the concrete with a hydraulic hammer attachment. A bucket and thumb or grapple is used to lift out the pieces. Cranes with wrecking balls are also used for the breaking operation. Under the right circumstances, explosives can be used to break up heavy foundations, which then can be removed with the excavator. Concrete foundations can also be broken up by using a concrete pulverizer attachment mounted on an excavator.

Occasionally, foundation concrete will be supported by piling, which can be wood, concrete, or steel. Usually, excavators with a bucket and thumb or grapple attachment can break the pile deep enough to make extraction unnecessary. An excavator with a shear attachment can also be used. If the piling must be totally removed, a crane or an excavator using a "vibratory extractor" attachment is most commonly used. This is discussed in more detail in chapter 9.

The following photos are examples of using the crane to "pick apart" various structures—*lifting down sections of buildings and structures*. This technique is often employed when certain building components are to be salvaged or the site is too confining to allow the structure to be felled either mechanically or by explosives.

Figure D.06. Felling elevated roadway by pulling.

Figure D.07. Pulling an industrial building.

E. SELECTIVE DEMOLITION

Selective demolition is defined as a careful demolition procedure whereby parts of a structure are removed while the primary structure is protected and remains intact. Selective demolition procedures use a combination of hand labor and small, specialized equipment. Examples of selective demolition are listed below:

- Removal of interior features such as walls, ceilings, and utilities—this is a very common procedure during remodeling work.
- Cutting openings in walls and floors for new utility services, stairwells, elevators, windows, doors, etc.

Figure D.08. Digging out foundations.

Figure D.09. Lifting down section of boiler.

Figure D.10. Tandem crane lift to remove two conveyor bridges. The larger conveyor bridge spanned 269 feet in length and weighed 200 tons. The two structures supporting the bridges were to remain intact and undamaged.

Figure D.11. Climbing scaffolding used for floor-by-floor demolition.

- Removing entire floors for major remodeling projects.
- Removing portions of structures that have become deteriorated or otherwise unusable.
- Removing portions of structures that have been damaged by fire, earthquakes, and similar events.
- Cutting and removing slots between building parts when one part of the building is to be demolished while the other part remains intact.

It is very important for the demolition contractor to carefully explain to his crew what is to be demolished and where the cut lines are located to ensure that his workers know exactly what is to be removed and what is to remain. The most effective way to ensure that the workers understand what is to be demolished and what must remain is to use a system of red and green spray paint markings. Some contractors use green marks to indicate what is to go (be demolished), while for others the green marks the construction features or equipment that are to remain and the red marks that which is to be removed. *It is important that everyone involved understands the marking system to be employed.*

Figure E.01. Mini-excavator with concrete cracker.

Typical equipment used for selective demolition work includes the following:

- Skid-steer loaders
- Small excavators
- Robotic excavators equipped with breakers and shears
- A wide variety of attachments for skid-steer loaders and mini excavators
- Aerial lifts and scissors lifts
- Scaffolding

Hand tools used in selective demolition include the following:

- Abrasive and reciprocating saws
- Pneumatic tools such as pavement breakers and chipping guns
- Concrete saws
- Chain saws

F. HAND DEMOLITION

Since the beginning of recorded history, man has used his hands and small tools to demolish structures and salvage useable materials. At some point during a mechanical or explosive demolition project, it is highly likely that small amounts of hand demolition work will be required.

Figure E.02. Selective demolition using a robotic machine.

Typical demolition activities that are performed primarily by hand include, but certainly are not limited to, the following:

- Cutting large steel sections that cannot be sheared or are located in areas that are inaccessible to heavy equipment
- Salvaging piping and valves
- Salvaging small equipment
- Salvaging electrical wiring, switch gear, and instrumentation
- Cutting access holes in walls and tanks
- Erecting and dismantling scaffolding and protective structures
- Removing many types of construction components and equipment for interior and selective demolition

Figure F.01. Using hand tools on floor with inadequate strength to support equipment.

G. SALVAGE

As mentioned in the Section C. of chapter 1, "Brief History of Demolition," salvage of building materials was the primary purpose for most of the early demolition efforts. The term "salvage" takes several forms and the more important types and methods of salvaging are discussed below.

The values of salvage sales to the demolition contractor can represent a significant portion of his or her income stream and allow increased competitiveness if he or she is reasonably accurate in the assessment of the salvage values for a particular job.

Equipment Salvage—industrial buildings particularly, and some modern commercial buildings, may contain useable equipment such as boilers, air conditioning systems, heat exchangers, tanks, electrical switchgear, motors, pumps, and wide variety of industrial process equipment. Usually, these types of salvage are carefully removed by the demolition contractor prior to beginning demolition operations.

Figure F.02. Building protection for gas meter.

Useable Materials Salvage—this category of salvage primarily includes timbers, dimensional lumber, piling, structural steel members, piping, and some types of brick.

Scrap Metals Salvage—by far, steel scrap produces the largest share of the salvage income for demolition contractors. Other important metals such as cast iron, copper, brass, aluminum, and lead are typically removed from the waste stream and sold to scrap recyclers.

Architectural Features—these items are sometimes referred to as "historic fabric" and include any part of a building that has value. Architectural features of certain buildings that no longer have adequate commercial value are sometimes salvaged for use in the façade or interior areas of new construction. Antique equipment may also be considered as historic and is sometimes salvaged for museum displays.

At times the issuing of a demolition permit is predicated on the demolition contractor agreeing to carefully salvage specific items considered to be of historic value.

H. HAZARDOUS MATERIAL MANAGEMENT

The proper management of hazardous materials is important to the successful operation of a demolition project. Although the demolition contractor may not have the contractual requirement to remove and dispose of hazardous materials, it is important that they are familiar with the hazardous waste management plan in force for each project.

Many of the larger demolition companies are qualified to perform hazardous materials work with their own forces, whereas other demolition firms will typically use companies that specialize in handling hazardous materials. The hazardous waste contractor may work directly for the project owner, as a subcontractor to the demolition contractor, or as a subcontractor to a general contractor. It is important for the demolition contractor to determine his contractual and legal responsibilities for identifying and handling hazardous materials before submitting a proposal for demolition to ei-

Figure G.01. Dismantling generator for sale.

Figure G.02. Copper wire collected for salvage.

Figure G.03. Salvaged valves.

ther an owner or general contractor. No matter who has the contractual responsibility for hazardous waste removal, every contractor working on a demolition project has the responsibility to protect their employees and the public from any hazards that exist on the project.

The federal government, through the U.S. Occupational Safety and Health Administration (OSHA), the U.S. Environmental Protection Agency (EPA), the U.S. Department of Transportation (DOT), and the U.S. Nuclear Regulatory Commission (NRC), has established extensive standards and regulations that must be followed during the handling of hazardous materials. Other federal agencies and most states also have enacted regulations that may add to the requirements for handling hazardous materials. These regulations and standards are covered in chapter 5. It is the intent of this section to introduce some common practices for hazardous waste management that may be encountered by the demolition contractor.

Identifying Hazardous Material on the Project—The Institute of Hazardous Materials Management (IHMM) provides the following definition of hazardous materials on their Web site: "A hazardous material is any item or agent (biological, chemical, physical) which has the potential to cause harm to humans, animals, or the environment, either by itself or through interaction with other factors" (http://www.ihmm.org/dspWhatIsHazMat.cfm). On the basis of this definition, the demolition contractor has a significant responsibility to protect its employees and the general public from any hazards that exist on the project, even if the hazardous material is not subject to regulatory enforcement. Protecting the safety of individuals and providing project management to shield the environment from long-term negative effects of demolition operations are important to company risk management.

What constitutes hazardous waste is not always clear from a regulatory standpoint. Regulations may exclude or include a material from a hazardous waste designation based on specifics of the physical or chemical use of the material in the building product, project conditions, removal techniques, or even if the material is to be recycled or reclaimed. State regulations may be different than federal regulations and are usually more restrictive.

Materials can be considered hazardous because they are listed as a hazardous material by standard or regulation, or simply because they exhibit characteristics that are hazardous. Characteristics that are frequently considered when determining if a material is hazardous include ignitability, corrosivity, reactivity, or toxicity. Some materials not typically subject to hazardous material regulatory enforcement that may present a liability to the contractor if not handled properly include used oils, lubricants, and other hydrocarbon-based products.

The following list includes some of the hazardous materials that may be found on a demolition project:

Asbestos Containing Materials (ACM): Friable ACM and Non-friable ACM	**Petroleum Oils Lubricants (POLs):** most POLs are not hazardous however removal and disposal is regulated
Polychlorinated Biphenyls (PCBs): require special removal, transportation, and disposal	**Radiological Hazards:** in some instrumentation and emergency lighting
Lead-Based Paints (LBPs): not all LBP is identified as hazardous, but testing is usually required before using a landfill for disposal	**Dust:** fugitive dust is usually non-hazardous, but may cause respiratory problems and contain fine particles of silica
Mercury: mostly from instrumentation, it is one of the most dangerous elements to humans	**Fluorocarbons:** refrigerants, these require special removal techniques, and some types may be recycled
Biological Hazards: bird and animal droppings, etc.	**Mold:** only some types of mold are hazardous
Chemicals: from industrial to household, not all chemicals are identified as hazardous	**Creosote:** railroad ties and pilings

In order to properly identify the presence of hazardous materials on a demolition project, a survey should be performed by personnel who are properly qualified by standards and regulations to investigate and assess the existence and quantities of potential hazards. This service is usually performed by specialized companies working directly for the owner. It is important for the demolition contractor to review hazardous materials surveys to be sure that the surveys meet professional standards. If surveys are not provided or appear to be incomplete, the demolition contractor may have to include the cost of surveys in his or her bid and make cost allowances for removal, transportation, and disposal as may be necessary according to the contract specifications. Surveys are usually performed by environmental contractors that are unrelated to either the owner or contractor.

Asbestos in some form is found on many projects built before the 1970s. The photos in the following section illustrate some removal and disposal practices commonly used for regulated hazardous material. The term Asbestos Containing Material (ACM) is used to describe materials containing a number of fibrous minerals that were used in the manufacturing of construction materials. Asbestos was a popular ingredient because it provided high heat resistance and strength. Unfortunately, breathing or ingesting the mineral fibers can lead to serious illness.

ACM is classified into the two main categories of friable and non-friable. Friable ACM is asbestos that can be crumbled or pulverized by hand pressure, releasing asbestos fibers. Non-friable ACM is asbestos containing material that does not easily release fibers when handled. The more hazardous friable ACM is commonly found in insulating materials and surface coatings. Floor and ceiling tile, roofing materials, and other firm or solid building products generally contain non-friable asbestos. *It should be understood that ACM normally considered to be non-friable can be made friable by aggressive removal procedures such as chipping, grinding, and pulverizing, or the forces of demolition.* When non-friable ACM is subjected to aggressive removal practices it generally triggers asbestos regulations not normally encountered for non-friable ACM.

A notification to EPA or its designated state agency of intent to remove ACM is required by National Emissions Standards for Hazardous Air Pollutants (NESHAPS) regulations. The notification must be filed at least ten business days before commencing removal work. Various states have notice and permit requirements and fees as well. The demolition contractor should make themselves aware of any state and local requirements prior to beginning demolition activity.

It is important to remember that during the demolition process unexpected and unidentified materials may be discovered. When unidentified material is discovered, the standard practice for a demolition contractor is to immediately cease work at the location of discovery. He or she must then initiate actions to obtain positive identification of such materials. If the identification results indicate a hazardous material, the contractor must implement action for its safe removal. In the case of discovery of hidden ACM (Figures H.02 and H.03), a new ten day notification may be required after the ACM is removed and prior to beginning demolition activity again unless an emergency permit is requested and granted.

Figure H.01. ACM prior to removal.

Care should be exercised when dealing with "empty" containers, tanks, or pipes. It is important for the contractor to physically inspect these items even if they have been assured that vessels have been emptied. In many cases a drain may have been opened, but due to project conditions, the tank or pipe still contains a significant quantity of material. Chemical processes may leave a residue that is hazardous or, due to the viscosity of the liquid contained in the vessel, some quantity remains adhered to the interior surfaces. Figures H.04 and H.05 show common examples. In Figure H.04, heavy oil was used as boiler fuel. The pipe drains were opened, but considerable oil remained in the pipe. Sections of pipe were cut and placed in a containment area in the warm sun. This allowed the oil to drain to an area where it could be collected for proper disposal.

Hazardous Material Removal—the removal process for hazardous materials requires a careful planning process and typically includes the following steps:

- Survey: The demolition site should be inspected by a qualified individual. This person prepares a survey report that describes the nature of the hazardous materials discovered, the location, and an estimate of the quantities. In some cases sampling and testing is required, and licensed personnel must complete the survey.

Figure H.02. ACM pipe insulation hidden in bathroom wall.

Figure H.03. Old and new ceiling, a common area of hidden ACM.

Figure H.04. Draining oil from "empty" pipe.

Figure H.05. Cleaning residue from "empty" tank.

- Notifications and Permits: Depending upon the geographic location, notifications to federal, state, and local authorities may be required.
- Training: There are a number of federal standards that require specific training for workers that will be involved with the removal and transportation of hazardous materials. Even if regulations do not specifically apply, training should be provided to prepare all workers for the hazard(s) anticipated.
- Personal Protection Equipment (PPE): All workers handling hazardous materials should be properly protected for the materials being handled. Care must be exercised to assure that the PPE utilized is specifically made to provide protection from the hazard(s) anticipated and fits the worker properly.

Figure H.06. Use of PPE for ACM removal.

Because asbestos containing materials are found on many demolition projects, the procedures employed when removing ACM can be used as an example of removal procedures for substances that are dangerous when inhaled or ingested. Hazardous materials that have different hazardous or physical properties would require modified procedures. For instance, the removal procedures required for liquid or gaseous hazards would need to be designed to protect against both the physical form of the material (liquid or gas) and the hazardous characteristic(s) (ignitable, corrosive, reactive, or toxic) of the material.

Contractors qualified to remove or abate ACM are required to possess specialized licenses. In addition, these asbestos abatement contractors employ supervisors and workers that have completed specialized training to become certified per OSHA standards. These training and certification programs are offered by providers specifically authorized to administer asbestos training.

As shown in Figures H.07 and H.08, separation must be maintained between the hazardous material removal process and any other activity in the area. An area of containment must be created to assure that the hazardous material does not spread into the general environment. All but workers properly trained and fitted with appropriate PPE must be excluded from the area (referred to as an "exclusion zone"). In the case of ACM removal, the containment must prevent airborne asbestos fibers from leaving the work area. This is generally provided by temporary disposable plastic sheeting. For some difficult to seal openings, expanding spray foam sealant may be utilized.

The containment area can be limited to the immediate work area or can include the enclosure of the entire structure. Due to the difficulty in completely sealing the containment area, the use of negative air pressure within the containment work space is usually required. By installing fans that draw air from within the containment space through high efficiency particulate air (HEPA) filters it is possible to make the air pressure within the containment space lower than the outside air pressure. This lowered (negative) air pressure forces any exchange of unfiltered air with the environment surrounding the containment to take place from outside to inside the containment. In this way, asbestos fibers within the containment area are prevented from reaching the atmosphere outside the containment. Figure H.09 shows an entire structure used as the containment. By observing how the plastic sheeting covering the windows is being sucked in, a basic indicator of negative pressure can be noted. Pressure monitoring equipment is typically installed to take note of the level of negative pressure achieved. Periodically throughout the day, pressure measurements are logged during ACM removal.

A containment area does not need to be large enough to contain workers, only the material that must be removed. Figure H.11 shows the use of glove bags to facilitate the removal of ACM. With the glove bag, workers are able to remove ACM within the bag without exposure to asbestos fibers.

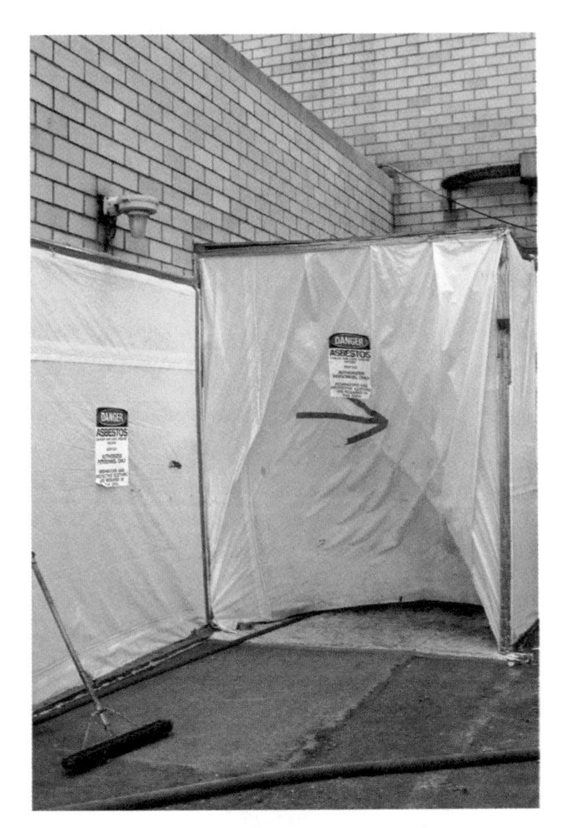

Figure H.07. Double flap entry to containment area. Note: negative air draws flap in.

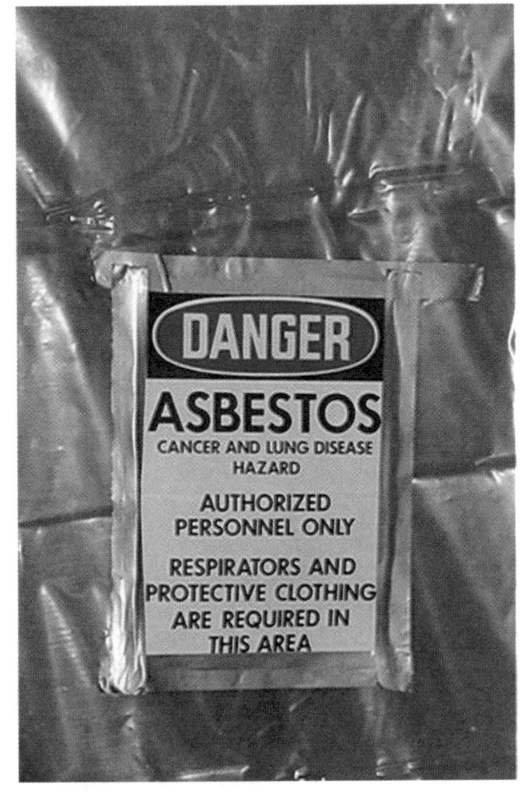

Figure H.08. Containment entry restrictions.

Figure H.09. Containment of entire structure. Negative air pressure shown on plastic window covering.

Figure H.10. Creating negative air pressure within containment.

Figure H.11. Glove bag for limited containment area.

Once the ACM is safely removed from the structure, it is left within the bag. At that point, the bag and its contents are disposed of with other ACM waste.

Removal of ACM is frequently a labor-intensive process. Water is used to prevent the asbestos fibers from becoming airborne. The wet ACM is placed in double strength plastic bags with appropriate warning labels for disposal (Figure H.12). All ACM waste containers leaving the site must be labeled to indicate the presence of asbestos within the container.

At times the ACM is repellant to the water. To assist with this problem a chemical surfactant is used as a wetting agent that lowers the surface tension of the water. The surfactant allows easier spreading of the liquid and more effective wetting of the ACM. After bulk removal of ACM is complete, HEPA filtered vacuums are sometimes used to remove asbestos fibers from restricted spaces (Figure H.13).

Hand removal of ACM may be supplemented by powered floor scrapers and power tools. Pneumatic tools are avoided to prevent the air released by the equipment from blowing asbestos fibers into the air. Compressed air is never used to clean surfaces or clothing for the same reason. Use of power tools is often limited by the need to decontaminate them after use to prevent the spread of asbestos fibers. Whenever possible disposable clothing and supplies are utilized and disposed of after use to prevent the release of asbestos fibers.

Regulations for removal of non-friable asbestos are less restrictive than friable asbestos and in some cases may be removed and disposed of as nonhazardous demolition debris as long as the removal process does not include cutting, chipping, grinding, or pulverizing that would release asbestos fibers. *ACM normally considered to be non-friable can be made friable by aggressive removal procedures or the forces of demolition.* When non-friable ACM is subjected to aggressive removal practices or other forces of demolition that may release asbestos fibers, it generally triggers asbestos regulations not normally encountered for non-friable ACM.

Some owners or general contractors will require third party monitoring during the asbestos

Figure H.12. Collecting ACM in labeled bag.

Figure H.13. Collecting ACM fibers with HEPA filter vacuum.

abatement process. An environmental consultant with no relationship to the abatement contractor is hired to monitor the hazardous material removal so that the possibility of future liability for environmental contamination is minimized. The monitoring program typically includes such activities as hazardous material testing and classification, emissions monitoring, hazardous jobsite access control, and project documentation.

Transporting Hazardous Materials—hauling hazardous materials requires compliance with the Department of Transportation (DOT) provisions specific to every state which the material is to pass through. These regulatory provisions include marking and labeling, use of proper containers, and provisions for reporting discharges of hazardous material while en route. Haulers must have appropriate state licenses and a valid EPA identification number for the transport of hazardous material.

It is also a good practice to require hazardous material haulers to provide an emergency response plan. The plan should include a listing of an emergency response vendor as well as a description of the hauler's own spill cleanup provisions. As in all demolition operations, the safety and training record of hazardous waste haulers is of interest to those contracting for these services.

ACM transport should take place in closed containers to prevent the release of asbestos fibers into the environment. Friable ACM debris is typically enclosed in double bags or otherwise contained such as in wrap and cut removal. When roll-off drop boxes are used to transport bagged ACM, a closed double plastic liner prevents the possible release of fibers during transport.

Hazardous waste must have chain-of-custody documentation that shows all parties that handle the material from jobsite to disposal facility. A Uniform Hazardous Waste Manifest is used for this purpose. This document (Figure H.16) is a standard EPA form that must be signed by a properly trained employee or designated agent as well as the waste transporter. A copy of the manifest must be mailed to the state (or states, if the disposal site is located in a different state than the project) within five business days. The original manifest must stay with the load through the full chain-of-custody, and a copy is retained by the contractor. When the load is received at the disposal site, both the transporter and the disposal facility's agent provide the final signatures. The transporter and disposal facility generally retain a copy of the manifest and mail the original to the contractor who submits a copy of the fully executed manifest to the state or states involved. If all compliance reporting is completed in a timely fashion, this process serves to provide assurance to all parties that the hazardous material was properly disposed of. Some hazardous materials are not regulated as hazardous waste and are not manifested. For instance, ACM transportation is documented with a waste shipping record.

Hazardous Material Disposal—All hazardous material must be disposed of at a facility designed and maintained to accept the specific hazardous material. In addition to documentation of material received through shipping records, hazardous material is typically subjected to visual inspection and/or testing at the disposal site. Disposal facility operators must characterize the content of each debris load to assure that it does not have the potential for environmental consequences that the disposal facility is not prepared to address.

Appropriate jobsite practices can help avoid debris load rejection due to shipment of hazardous material that does not have a profile appropriate for the disposal facility. At the least, good jobsite hazardous debris practices will minimize unnecessary testing when the load reaches the disposal facility and lower disposal costs by allowing debris to be shipped to the lowest cost disposal alternative.

By segregating materials at the jobsite according to their characteristics, only hazardous material will be shipped, and mixed loads are avoided. Segregation also promotes lowered disposal cost by allowing optimum debris disposal for different debris types. Storage of hazardous materials on site should be avoided. As with all debris, this practice avoids costly double handling of material. It also helps avoid errors in labeling or handling of containers.

Some hazardous materials can be recycled. This is especially true of the various hazardous materials that may be found in small quantities in some building products. Examples include mercury in fluorescent tubes, chemical compounds in batteries, small radioactive sources in smoke detec-

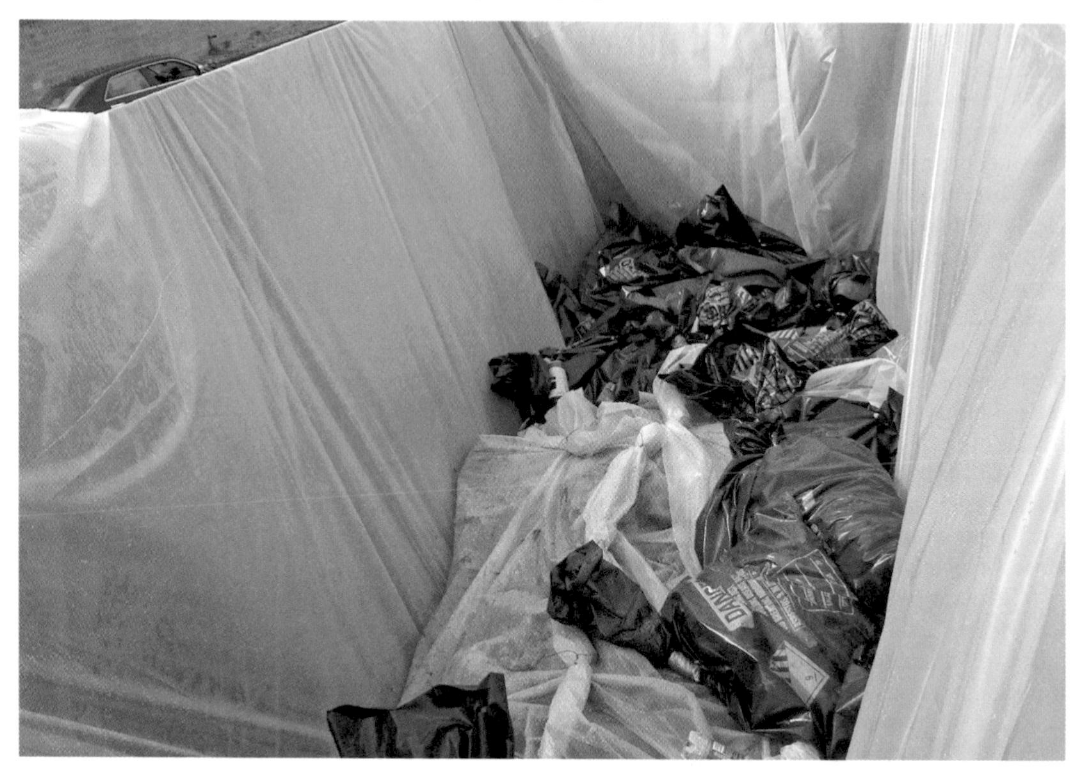

Figure H.14. ACM transport container with double layer of plastic film to provide added protective closure.

Figure H.15. Transport container accessible to building containment.

Please print or type. (Form designed for use on elite (12-pitch) typewriter.) Form Approved. OMB No. 2050-0039

UNIFORM HAZARDOUS WASTE MANIFEST | 1. Generator ID Number | 2. Page 1 of | 3. Emergency Response Phone | **4. Manifest Tracking Number**

5. Generator's Name and Mailing Address | Generator's Site Address (if different than mailing address)

Generator's Phone:

6. Transporter 1 Company Name | U.S. EPA ID Number

7. Transporter 2 Company Name | U.S. EPA ID Number

8. Designated Facility Name and Site Address | U.S. EPA ID Number

Facility's Phone:

9a. HM	9b. U.S. DOT Description (including Proper Shipping Name, Hazard Class, ID Number, and Packing Group (if any))	10. Containers No.	10. Containers Type	11. Total Quantity	12. Unit Wt./o l.	13. Waste Codes
	1.					
	2.					
	3.					
	4.					

14. Special Handling Instructions and Additional Information

15. **GENERATOR'S/OFFEROR'S CERTIFICATION:** I hereby declare that the contents of this consignment are fully and accurately described above by the proper shipping name, and are classified, packaged, marked and labeled/placarded, and are in all respects in proper condition for transport according to applicable international and national governmental regulations. If export shipment and I am the Primary Exporter, I certify that the contents of this consignment conform to the terms of the attached EPA Acknowledgment of Consent.
I certify that the waste minimization statement identified in 40 CFR 262.27(a) (if I am a large quantity generator) or (b) (if I am a small quantity generator) is true.

Generator's/Offeror's Printed/Typed Name | Signature | Month Day Year

GENERATOR

16. International Shipments ☐ Import to U.S. ☐ Export from U.S. Port of entry/exit:
Transporter signature (for exports only): Date leaving U.S.:

INT'L

17. Transporter Acknowledgment of Receipt of Materials
Transporter 1 Printed/Typed Name | Signature | Month Day Year
Transporter 2 Printed/Typed Name | Signature | Month Day Year

TR ANSPORTER

18. Discrepancy
18a. Discrepancy Indication Space ☐ Quantity ☐ Type ☐ Residue ☐ Partial Rejection ☐ Full Rejection
Manifest Reference Number:
18b. Alternate Facility (or Generator) | U.S. EPA ID Number
Facility's Phone:
18c. Signature of Alternate Facility (or Generator) | Month Day Year

19. Hazardous Waste Report Management Method Codes (i.e., codes for hazardous waste treatment, disposal, and recycling systems)
1. | 2. | 3. | 4.

20. Designated Facility Owner or Operator: Certification of receipt of hazardous materials covered by the manifest except as noted in Item 18a
Printed/Typed Name | Signature | Month Day Year

DESIGNATED FACILITY

EPA Form 8700-22 (Rev. 3-05) Previous editions are obsolete. DESIGNATED FACILITY TO DESTINATION STATE (IF REQUIRED)

VOID

Figure H.16. EPA Uniform Hazardous Waste Manifest.

Figure H.17. Crushing fluorescent tubes for recycling.

tors, refrigerants, and mercury in thermostats and switch gear. Each of these individual components contains a small quantity of hazardous material that is magnified when a large number of individual components are collected. Figure H.17 shows the use of equipment to crush and collect the debris from fluorescent tubes in containers appropriate for shipment to a recycling facility.

I. STUDY QUESTIONS

1. Describe the method you would use to demolish the following structures; include the type of equipment and any special considerations you think would be necessary:

   a) Two story apartment building with good access
   b) An eighteen story office building within fifty feet of four story buildings on two sides
   c) A 200 foot antenna located on a ten acre field with no other structures on the property
   d) A six story brick and timber warehouse that has suffered partial collapse
   e) Salvage of a ten ton vessel from a 120 foot high industrial building
   f) Salvage of an historic stained glass window from a church
   g) Removing the brick liner from an industrial kiln that is contaminated with caustic residues

2. What equipment could be used to break up a 4 foot thick concrete foundation?

3. What are four reasons for not using explosives to bring down a 150 foot high water tank?

4. Why is salvaging metals so important to a demolition project?

5. Describe what qualities when present in a material would lead a demolition contractor to consider the material to be hazardous.

6. List twelve hazardous materials commonly found on a demolition project.

7. Describes the steps typically undertaken by demolition contractors when planning for removal of hazardous material.

8. Describe three procedures used by demolition contractors to protect the general public during hazardous material removal.

9. What is the difference between friable and non-friable asbestos?

10. Name some hazardous materials that can be recycled.

11. Answer the following questions about chain-of-custody documentation.

    a) What is chain-of-custody documentation?
    b) When and why is it required?
    c) Who regulates the documentation process?
    d) What standard form is used to monitor the documentation process?

CHAPTER 4

TYPES OF DEMOLITION—BUILDINGS AND STRUCTURES

A. GENERAL

This chapter reviews the various types of buildings and structures that are commonly demolished by demolition contractors. It is important to understand that even though some projects are relatively simple, all demolition must be performed with safety as the primary consideration both in selecting the equipment to be used and the demolition technique to be employed.

At this time, the authors advise the reader that even though some projects look simple, such as a single family house compared to a high-rise building, there is always the problem of complacency. Both key management personnel and workers sometimes become inattentive when the tasks being performed are very routine and simple, leading to mistakes and accidents. Regardless of the size or complexity of a project, the demolition contractor must ensure that all the demolition work plan tasks and safety requirements are strictly followed.

Another factor in the demolition of buildings is cleanup. The actual demolition and hauling is relatively straightforward, but picking up the "sticks" and miscellaneous work items can take considerable time and result in significant costs. Do not count the profits until all the sticks are gone and the owner signs off on the job as complete.

Often even on a relatively small demolition project, the real work and cost begins after the building has been demolished and foundation removal begins. Whereas the above grade structure can usually be assessed and measured, the below grade foundations can seldom be inspected. Chapter 6 discusses the methods for determining quantities of foundations.

B. LOW-RISE COMMERCIAL AND RESIDIENTIAL

Typically the demolition of one- to three-story buildings is the least complex type of project. The three-story height of thirty-six feet or less is easily accessed by twenty-five to fifty ton excavators, and most companies in the demolition business can do this type of work. For one- and two-story buildings, such as housing, the front-end track loader is commonly used to wreck the building, grind the debris, and load the trailers or drop boxes. Smaller demolition firms are usually very competitive on this type of demolition.

C. HIGH-RISE BUILDINGS AND OTHER STRUCTURES

This category of demolition includes a very significant percentage of the total work done by more experienced demolition contractors.

D. FOUNDATIONS

All structures, from buildings to bridges, are supported by foundations. Foundations come in many forms and are often the most costly part of a demolition project. Common foundation types are listed below:

- Buildings and other structures—foundation walls, spread, and column footings
- Machinery foundations
- Basement walls and footings
- Retaining wall foundations and spread footings
- Bridge piers and abutments

Figure B.01. Demolition of low-rise industrial building.

Figure C.01. Stadium implosion.

Figure C.02. Building demolished floor-by-floor.

Figure C.03. Missile launch tower being disassembled.

Figure D.01. Breaking concrete foundations.

- Smokestack foundations
- Massive "floating" foundation pads
- Tower and antenna foundations
- Foundation pilings—either wood, steel, or concrete
- Grade beams

E. BRIDGE DEMOLITION

Demolition of bridges in North America has become another significant category of the demolition industry. Infrastructures, including bridges of all types, are aging, and literally thousands of sub-standard bridges are to be replaced in the early decades of the twenty-first century. Most demolition contractors bid bridge demolition work along with the wide variety of other types of work; however, there are a few firms that tend to specialize in bridge demolition work. There are several bridge categories such as highway bridges, railroad bridges, pedestrian bridges, and utility support bridges. They all come in different forms and all of them present unique challenges.

Bridges can be constructed of concrete, steel, wood, and composites of two or more materials. The following photos are examples of larger highway and railroad bridges.

Figure E.01. Excavator shearing bridge girders.

Figure E.02. Hand labor demolishing bridge deck with pneumatic tools.

Figure E.03. Removal of bridge girders with hydraulic hammer.

The demolition of a bridge can be a very complex operation requiring extensive advance planning and coordination. For example, the planning and operational tasks of a bridge demolition may include the following:

- Permits—often several different types of permits are required. Typical permits are listed below:
    - permit for traffic revisions
    - permit for working over a waterway
    - permit for working in or close to a wetland
    - permit for night work, if required
    - permit from the Coast Guard to work or obstruct a navigable waterway
    - permit from railroad company to work on their property
- A comprehensive engineering study of the condition of the structure may be required, particularly if there is visible structural deterioration.
- A detailed engineering plan for demolishing the bridge superstructure and foundations. The plan may also include constructing and removing any temporary shoring that may be required during the process of demolition. Working over or in close proximity to live traffic is inherently dangerous, and the demolition contractor must exercise extreme caution.
- Coordination of operations with departments of transportation and municipal government agencies. Railroad companies and the U.S. Coast Guard may also be involved in some projects.

Figure E.04. Excavator with concrete pulverizer demolishing highway bridge.

Figure E.05. Bridge deck removal with hydraulic impactors.

F. INDUSTRIAL DEMOLITION

Industrial demolition can be loosely defined as any type of demolition that takes place in an industrial or factory setting. Typical industrial demolition projects are listed below with photographs following:

- Refineries
- Pulp and paper mills
- Steel mills
- Chemical plants
- Food processing plants
- Pharmaceutical companies
- Electronics manufacturing companies
- Power plants
- Industrial manufacturing facilities

Note from figures F.01, F.02, and F.03 that the excavator and shear attachments are the primary pieces of demolition equipment for industrial demolition. Cranes also play an important role in the demolition of industrial structures.

Many industrial demolition projects involve partial removals of plant structures and process systems while the facility remains in operation. For example, an obsolete refinery cracking tower might be demolished in an area that has active gasoline product pipelines. These projects require careful planning and coordination with the owners' representatives and the implementation of very strict safety rules.

Figure F.01. Shearing petroleum tanks with excavator.

Figure F.02. Demolition with excavator and shear.

Figure F.03. Pulling over a stack.

G. MARINE DEMOLITION

Marine demolition may be loosely defined as "demolition of any type of structure that requires operation on or adjacent to a body of water." Access to marine demolition projects may be from the shore or from barges. Divers may be needed in certain circumstances, and explosives are sometimes used for bridge demolition and underwater breaking of concrete foundations. In general, marine structures may include any of the following:

- Bridges—both superstructures and foundations, over waterways
- Piers, docks, or wharves
- Piling
- Material handling systems
- Sunken boats or barges
- Submerged pipelines

Bridges—many bridges are over water and may require work from barges as well as access from shores. A typical bridge demolition over a body of water usually occurs after a new bridge is constructed and takes the traffic load. This situation requires considerable engineering planning in order to ensure there is no operational disruption of traffic on the new bridge.

Piers, Docks, or Wharves—a pier is an over water structure attached to the land and supported by piling. Docks or wharves are built along shorelines. These structures have many uses such as docking areas for ships and barges, support for buildings, and amusement parks. Many of these demolition projects are the result of damage caused by fire or storm and deterioration due to rotting or rusting.

Figure G.01. Excavator on barge demolishing concrete bridge pier with shear.

Figure G.02. Removing piling from rotten pier.

Piling—piling is the most common form of base support for structures in the water. Pile can be wood, concrete, or steel. Wood pile may be either completely pulled out or, more commonly, broken off below the mud line so there will not be any pile stubs that could be dangerous to navigation. Steel and concrete pile are usually either cut with a shear or completely pulled out because of the difficulty of breaking them. To extract piling, a vibratory pile extractor is commonly used. Landfill disposal of wood pile treated with creosote is regulated in the United States.

Material Handling Systems—such systems include over water conveyors for loading ships with grain, coal, scrap steel, and a host of bulk materials. The conveyors may be supported by either pile foundations or are built on existing wharves or piers.

Sunken Ships and Barges—the demolition contractor is rarely called upon to remove sunken vessels, but it does occur. Such work can be accomplished with a crane or an excavator. Work may be from the shore or from a barge.

Figure G.03. Dry dock demolition.

High capacity, barge mounted cranes are ideally suited for most types of over water demolition projects and are usually rented when the demolition contractor has a need for heavy lifts. Mobile or crawler cranes and excavators can be loaded on barges and secured to the deck for some types of marine projects.

H. HISTORIC SALVAGING

Historic salvaging is the careful removal of building components for their historic value. There is no limit to the type of building features that may be classified as historic and considered for salvage. Many of these items wind up in museums, parks, and private or public buildings. Examples of the more common historic material salvaged:

- Special bricks and terra-cotta
- Carved stonework
- Building entries
- Stained glass windows
- Interior woodwork of all kinds
- Cast iron features
- Bells and belfries
- Timber pieces from historic buildings
- Antique machinery

Figure H.01. Historic terra-cotta figurehead to be salvaged.

There are some projects involving historic salvage that require the careful removal of entire building facades for reconstruction elsewhere. This can be accomplished by a well-planned system of photography and match marking. The demolition contractor may have his own skilled personnel for such work; however, for very special items, such as stained glass, the contractor may hire a firm specializing in such work.

Figure H.02. Careful removal of a historic cupola for reinstallation.

I. SPECIAL STRUCTURES

Special structures are any kind of structure outside of the normal classifications of buildings or bridges. Stand-alone structures and many industrial projects will require the demolition of highly specialized process equipment and/or structures. The following photographs are examples of a few special structures encountered by demolition contractors:

J. DEMOLITION OF PRE-STRESSED AND POST-TENSIONED CONCRETE STRUCTURES

The demolition of pre-stressed concrete structures presents the demolition contractor with unique problems. Pre-stressing of concrete is accomplished by tensioning special steel cables within the concrete mass. There are two primary methods for pre-stressing concrete structures: (1) Pre-tension-

Figure I.01. Pulling a radar dome.

Figure I.02. Dismantling missile launch structure.

Figure I.03. Ring platform for demolishing stack.

ing: by the application of tensioning forces before the concrete is cured so that the friction between the cables and the concrete is strong enough to maintain the tension desired, or (2) Post-tensioning: where metal or plastic tubes that contain loose tensioning cables or rods are cast into the concrete and when the concrete reaches a certain design strength, the cables are then tensioned to a pre-determined force. In both cases, the ends of the cables or rods are anchored to each end of the building component by various types of devices.

It is extremely important for the demolition contractor to determine the presence of post-tensioned concrete construction before demolition operations begin. Pre-tensioned construction is usually not critical from a demolition standpoint. The determination of either pre-tensioned or post-tensioned construction should be made during the Engineering Survey by a competent person. In performing the Engineering Survey, the competent person can usually detect the presence of pre-stressing by closely examining the various building or bridge components such as beams, columns, and slabs. Drawings of the structure should also be reviewed to determine the presence of pre-stressing. If there is any doubt as to whether pre-stressing is present, the demolition contractor should obtain the advice of an engineer specializing in pre-stressed construction.

For structures such as buildings and parking garages, built with pre-tensioned members, the demolition process is virtually the same as for cast-in-place construction. Since the tensioned cables are bound to the concrete they do not tend to violently de-energize when the concrete is broken and the tension is released. For selective demolition, undamaged pre-tensioned members can also be cut loose from their attachments to the rest of the structure and lowered down with a crane if necessary.

Post-tensioning is a different matter. Since the cables or rods are not bound to the concrete, they can act like a stretched rubber band. When the concrete is broken the cable may snap violently, causing the embedded anchors to become missiles. In addition to the potential for anchors to become ejected, the sudden release of the tendons causes the post-tensioned member to lose its tensile strength, causing the concrete to fail. The result can be a catastrophic collapse of the entire building.

For post-tensioned structures that can be imploded, the risk of injury or damage is controlled by ensuring that people are kept at a safe distance and that there is no exposure to property if anchors are ejected. To restrain anchors from being ejected, sandbags or other measures are utilized to temporarily hold the anchors during collapse.

For post-tensioned structures that can be demolished using a crane and ball or high-reach demolition equipment, the system for progressive demolition can be designed to allow the machine to work at a safe distance from the structure should collapse occur. Temporary anchor restraints may also be needed.

It is advisable for the demolition plan describing the demolition of a post-tension building in a confined location, such as a downtown area, to be designed by an engineer experienced with post-tensioning construction. Usually, the building must be supported by shoring from the basement to the roof. This allows for a floor-by-floor demolition method to be used using hand labor and small machines. Overloading of the floors can be eliminated by removing the demolished materials without delay.

Note: The subject of demolishing post-tension structures can get very technical, and this brief summary does not address all of the issues that must be taken into account.

K. INTERIOR DEMOLITION

Interior Demolition is usually the first step in preparing a building for remodeling and represents a significant segment of the demolition industry. This type of work covers everything from removing a few interior partition walls to total removal of all non-structural building components. A number of demolition firms have specialized in interior demolition and utilize a wide variety of small equipment and tools to accomplish this work. A simple example of interior demolition is thê removal of floor coverings, partitions, and ceilings from a small slab-on-grade building using skid steer loaders equipped with grapple buckets and hand labor. A more complex example may involve partial demolition of an occupied multi-story building, which requires much planning and often may require

Figure J.01. Pre-stressed concrete beams and columns.

Figure J.02. Post-tensioning tendons with anchor head removed.

night work to avoid disrupting tenants. In some cases shoring will be required to safely distribute floor loads and allow the use of small equipment.

Figure K.01. Debris chute for interior demolition.

Figure K.02. Building wrapped.

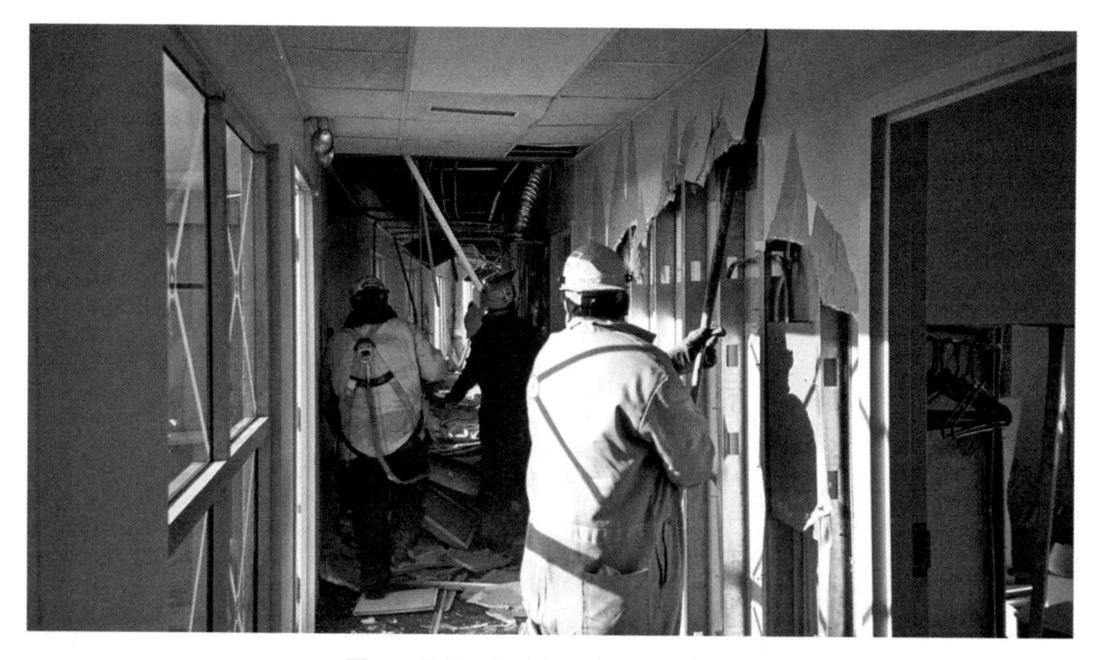

Figure K.03. Hand demolition of interior.

Figure K.04. Small loader for interior demolition.

L. STUDY QUESTIONS

1. What is widely considered the most costly part of a demolition project?

2. List two of the highly specialized work plans often required on demolition projects.

3. In the complicated process of bridge demolition, what is commonly required if there is visible structural deterioration?

4. Name a few of what are considered typical industrial demolition project types.

5. What is the main reason piers, docks, or wharves are demolished?

6. Describe the process of pre-tensioning and post-tensioning concrete. What impact does the presence of each type of concrete reinforcement have on a demolition project?

REFERENCES

National Demolition Association. 2004. *Demolition Safety Manual*. Doylestown, PA: NDA.

CHAPTER 5

DEMOLITION REGULATORY GUIDES

A. INTRODUCTION

We live in a highly regulated society and a variety of regulations apply to the demolition industry. The regulations and standards applicable to demolition work are an important part of overall safety in the demolition industry. Regulations commonly associated with demolition work can be classified in the following categories:

- Safety
- Environmental—site requirements
- Environmental—hazardous materials requirements
- Community-specific regulations

Each of these categories contains numerous federal, state, and local standards that play an important role in the means and methods of how demolition work is accomplished. These standards should be considered as a guide to performing all types of demolition safely. Even when complying with these standards, it is important to remember that they are in no way a substitute for common sense.

Generally speaking, these regulations are designed to prevent injury to people or contamination of the environment. Most were initially a response to an accident, hazardous material spill, or common practice that new knowledge or study found to be dangerous to the general health and welfare of the public.

The numerous regulatory standards referred to herein should be referenced on an as-needed basis when determining the regulations that are applicable to a particular project. Although most share a common origin in U.S. law, municipal standards often deal with local concerns that should not be overlooked.

All states are required to mandate compliance to at least the level of performance stipulated by federal standards, but have the ability to adopt rules more stringent than the federal standards. Similarly, local governmental bodies such as counties and cities can adopt rules more stringent than the state standards.

Many project specifications will list references to various standards as they may apply to a particular project. If one is unfamiliar with a listed standard, it is important to read it and verify that it pertains to the work to be done. The standards must in turn be enforced by those supervising and working on the project.

B. OCCUPATIONAL SAFETY AND HEALTH ADMINISTRATION (OSHA)

For the protection of all workers in the construction industry, the federal Occupational Safety and Health Act became law in 1970 establishing the Occupational Safety and Health Administration (OSHA). OSHA and the standards the agency has promulgated have been incorporated by the National Demolition Association (NDA) into a *Demolition Safety Manual* that covers the typical activities to be expected on a demolition project. Table 5.01 below is a list of the OSHA standards addressed in the NDA Manual. These constitute important safety tasks designed to protect the employees working on a demolition project. It is important to note that the OSHA standards carry the force of federal law. Companies and individuals that violate OSHA standards can, and do, face costly fines. In some cases the responsible individuals can be tried and sentenced to jail terms for willful violations.

Table 5.01

General Category	OSHA Reference	Description of Activity
	29 CFR Applies to all	
General Safety and Health	1926.20	Integrating Safety Into the Job
	1926.21	Importance of Training and Education
	1926.23	Medical Services and First Aid
	1926.25	Fire Prevention and Protection Plan
	1926.26	Signs and Lighting
	1926.56	Ventilation
	1926.150 -155	Fire Prevention and Protection
	1926.200	Signs and Tags
Personal Protective Equipment	1926.28	Safe Work Clothing, Hand Protection
	1926.96	Foot Protection
	1926.100	Head Protection
	1926.102	Eye and Face Protection
	1926.101	Hearing Protection
	1926.103, 1910.134	Respiratory Protection
	1926.104	Safety Belts, Lifelines, and Lanyards
Subpart T Demolition	1926.850 -1926.860	Preparatory Operations
	1926.850(a)	Engineering Survey
	1926.850(c)	Utility Location
	1926.852	Debris Removal
Fall Protection	1926.500	Scope, Application, and Definitions
	1926.501	Duty to Have Fall Protection
	1926.502	Fall Protection Systems Criteria and Practices
	1926.502(i)	Hole Covers
	1926.503	Training Requirements
Stairways and Ladders	1926.1050 - 1053	
Scaffolds	1926.450 - 452	
	1926.453	Aerial Lifts
	1926.454	Training
	1926.550(g)	Suspended Personnel Platforms
Safe Use of Hand Tools	1926.300	General Requirements
	1926.301	Hand Tools
	1926.302	Power Tools
Welding and Torch Cutting	1926.350	Gas Welding and Cutting
	1926.352	Fire Prevention
	1926.353	Ventilation and Protection
	1926.354	Preservative Coatings
	1926.351	Arc Welding and Cutting
Confined Space	1926.21.b.6	Safety When Working in Confined Spaces
	1910.146	Permit Required Confined Space
Handling Hazardous Materials	1926.1101	Safe Handling of Asbestos
	40 CFR Part 763	Safe Handling of Asbestos
	40 CFR Part 761	Safe Handling of PCBs
	1926.62	Medical Surveillance Guidelines
Cranes and Derricks	1926.550 .a1 - .g8	
Equipment Safety	1926.600	Equipment
	1926.601	Motor Vehicles
	1926.601(b)(14)	Ensuring Safe Operating Equipment
	1926.602	Material Handling Equipment
		Safe Use of Material Handling Equipment
Mobilization		Preparing Machinery for Transport
Protective Structures	1926.1050	Shoring
Demolishing Chimney & Stacks	1926.854	Safe Work Practices

Pre-Stressed Concrete	Note: See chapter 8 for detailed discussion	Demolition of Pre-Stressed Concrete Structures
Safe Blasting Procedures	**1926.900**	Safe Blasting Procedures—General Provisions
	1926.900(d)	Storage of Explosives
	1926.904	Storage of Explosives and Blasting Agents
	1926.914	Definitions Applicable to Explosives
	1926.900(b),(g)	Transport of Explosives
	1926.902	Surface Transportation of Explosives
	1926.903	Underground Transportation of Explosives
	1926.901	Blaster Qualifications
	1926.905 through	Loading of Explosives or Blasting Agents
Hazard Communication	**29 CFR 1910.1200**	Refer to the Entire Regulation for Requirements of 29 CFR 1910.1200
Competent Person	**29 CFR 1926.32(f)**	Defines "Competent Person"—means one who is capable of identifying existing and predictable hazards in the surroundings or working conditions

The material for the above Table 5.01 was taken from *Demolition Safety Manual*, 2004 edition, and the "Hazardous Communications Program," both of which were prepared by the National Association of Demolition Contractors (now known as the "National Demolition Association").

C. AMERICAN NATIONAL STANDARDS INSTITUTE (ANSI)

Guidelines for the safe performance of demolition and related activities are published by the American National Standards Institute in "Safety and Health Program Requirements for Demolition Operations" Reference ANSI/ASSE A10.6.

The ANSI standards A10.6 are a guide for use by government agencies and other interested parties to provide a system to convey to contractors and their employees the suggested safe practices to be followed during demolition operations and related activities. It is to be noted that ANSI was the foundation for much of the OSHA standards, and it is an excellent source of information regarding all forms of construction safety. The ANSI standards do not carry the force of law but may be referenced in contract requirements.

D. U.S. ARMY CORPS OF ENGINEERS SAFETY AND HEALTH REQUIREMENTS MANUAL, EM 385-1-1

This safety and health manual has been in use for many decades as a standard for many federal government projects and all U.S. Army projects. Section 23, "Demolition," addresses demolition operations specifically; other sections deal with general construction activities, many of which are commonly related to demolition operations. For example, Appendix A of EM 385-1-1 provides a "Minimum Basic Outline for an Accident Prevention Plan," and Section 29 "Blasting" covers most activities that may be expected on a demolition project where explosives are to be used. Also of particular importance are Section 15, "Rigging," and Section 16, (C) and (D), which deal with cranes and their safe operation.

E. CODE OF FEDERAL REGULATIONS (CFR)

The CFRs comprise many hundreds of pages of regulations that have evolved over the years to deal with most all facets of the federal government's activities. The CFRs are a part of the Federal Acquisitions Regulations (FARs). For example, the FARs set forth specific rules from FAR Subchapter E, Part 28, Bonds and Insurance, to Subchapter D, Part 23, Subpart 23.3, Hazardous Materials Identification and Material Safety Data. All of the OSHA regulations discussed above are important to the demolition industry. The regulations titled 29 CFR 1926, Subpart T Demolition, constitute those OSHA regulations that are specifically directed to the demolition industry.

Although 29 CFR 1926, Subpart T Demolition, constitute the OSHA standards that are specifically directed to the demolition industry, there are many more standards and regulations that may apply to a demolition project. The regulations discussed above and some of the OSHA standards, from 29 CFR 1926 Construction, 29 CFR 1910 General Industry, and 29 CFR 1915 Maritime can apply to a demolition project.

Other FARs applicable to most demolition work are listed below:

- FAR 29 CFR 1910 Occupational Safety and Health Standards
- FAR 40 CFR 122 National Pollution Discharge Elimination System (NPDES)
- FAR 40 CFR 260-265 Resource Conservation and Recovery Act (RCRA)
- FAR 40 CFR 300-399 Comprehensive Environmental Response, Compensation, and Liability Act (CERCLA)
- FAR 40 CFR 700-789 Toxic Substance Control Act (TSCA)
- FAR 40 CFR 761-763 Asbestos
- FAR 40 Subpart M, National Emission Standards for Hazardous Air Pollutants (NESHAP)

F. ENVIRONMENTAL REGULATIONS

Environmental regulations may be classified as "Environmental Site Requirements" and "Hazardous Materials Requirements." Many regulations address environmental protection concerns and are issued by federal, state, and municipal regulatory authorities. Examples are listed below.

Environmental Site Requirements:

- The National Pollution Discharge Elimination System (NPDES) administered by the EPA requires that the owner of a demolition site over one acre obtain a permit and devise and implement a plan to prevent pollutants conveyed by rainwater or snow melt from leaving the site. This is currently one of the most cited violations by enforcement authorities. For demolition projects that cover an area of one acre or more, a Storm Water Pollution Protection Plan (SWPPP) must be prepared. This plan will describe the means and methods for controlling runoff from rainwater and snow melt and will typically be prepared by persons that specialize in such plans. A simple SWPPP may require engineering controls such as silt fences and methods to protect storm drains from silt or other contaminants contained in runoff. In some cases, it may be necessary to collect water runoff and filter it to standards of purity defined by regulations. The purpose of these regulations is to protect nearby bodies of water from siltation and contamination from hazardous materials that may be present in the soil. In many cases, the owner of the property or a general contractor may already have a SWPPP in effect, and the demolition contractor will be required to comply with its requirements.
- FAR 40 CFR Subpart M, National Emission Standards for Hazardous Air Pollutants (NESHAP) covers the air quality requirements that may be expected of most demolition operations. Among other things, this regulation requires a notification postmarked or delivered no later than ten working days prior to the beginning of asbestos removal or demolition projects. Most states and municipalities also have regulations governing the engineering controls required to limit the amount of dust on a demolition project.
- Noise control—exposure to noise in the workplace is governed by OSHA 29 CFR 1926.52, which established noise exposure levels in terms of hours per day at various decibel levels. The EPA has

allowed states to establish noise levels for many types of construction-related activities. Some municipalities have noise ordinances to limit the noise level or the time of day that certain noise levels are allowed in the community.

Figure F.01. Protecting storm drain.

Hazardous Materials Requirements:

There are extensive environmental regulations relating to the identification, removal, transportation, and disposal of hazardous materials. In fact, the costs of removing and disposing of hazardous materials can sometimes be more than the cost of the demolition process itself. Before demolition operations can begin, it is necessary to thoroughly inspect the building and identify any suspected hazardous materials.

Hazardous materials are then removed if required by regulation. Even if not required by regulation, removal and segregation at this point may result in more cost effective disposal of the hazardous debris. If hazardous materials are found during the process of demolition, the contractor must cease work in the affected area, take proper steps to confirm identification of such materials, and as necessary allow removal and disposal of them as prescribed by regulations.

There are three important federal statutes that address the problem of hazardous materials in the workplace and the general environment:

- CERCLA—Commonly referred to as the "Superfund" is set forth in The Comprehensive Environmental Response, Compensation, and Liability Act (CERCLA). CERCLA was enacted by Congress to create a tax on the chemical and petroleum industries. CERCLA provides broad federal authority to respond directly to releases or threatened releases of hazardous substances resulting from closed and abandoned hazardous waste sites that may endanger public health or the environment. It also established a trust fund to provide for cleanup when no responsible party can be identified. The law authorizes short and long term remedial actions to address hazardous waste releases at these sites. These actions can be conducted only at sites listed on EPA's National Priorities List (NPL).

- RCRA—The Resource Conservation and Recovery Act is designed to protect human health and environment by establishing a comprehensive regulatory framework for investigating and addressing past, present, and in some cases future environmental contamination at hazardous waste treatment, storage, and disposal facilities. RCRA designates how a listed waste is categorized, based on criteria such as reactivity or toxicity. A demolition waste that contains a RCRA listed material, such as cadmium or arsenic, may be classified as either construction and demolition debris or as a hazardous waste, depending on analysis of the leachable levels of the RCRA material. These factors can have significant disposal cost impact.
- NESHAP—National Emission Standards for Hazardous Air Pollutants regulations cover a wide variety of potentially harmful air pollutants. The NESHAP regulation of asbestos containing material is of particular importance to the demolition industry.

Listed below are some hazardous materials that may be encountered on a demolition project:

Asbestos Containing Materials (ACMs) are present in some form on many demolition jobs. ACM comes in many forms and degrees of hazard. Both OSHA and the EPA, along with most states, have very thorough regulations stipulating how the ACMs are to be handled, transported, and disposed of. It is very important to provide the proper (typically ten day) notification to the appropriate regulatory agencies when asbestos removal is required.

Polychlorinated Biphenyls (PCBs) are commonly found in an assortment of petroleum products, paints, and other products such as electrical equipment. Disposal of PCB waste and containers or PCB contaminated soil and concrete are highly regulated.

Lead-Based Paints (LBPs) are frequently found on building construction materials, and demolition of LBP coated material must comply with the OSHA 1926.62 Lead Standard. In a building undergoing partial demolition, even undisturbed LBP may have to be removed depending on the location and future use of the building. In most cases, LBP debris can be handled as a normal part of the waste stream if the Toxicity Characteristic Leaching Procedure (TCLP) tests show it to be below the Maximum Contamination Levels (MCLs). However, if the LBP debris is above the MCLs, it may have to be handled as hazardous material.

Mercury is found in a variety of instruments such as thermostats, pressure sensors, and as a vapor in fluorescent lights. Mercury is extremely hazardous to human health and must be removed and handled with great care. Environmental releases and disposal of mercury are regulated primarily by the EPA.

Petroleum Oils and Lubricants (POLs) are usually not a hazardous material but they must be removed and properly disposed of wherever they are encountered on a demolition project. The removal of both underground and above ground oil storage tanks are common tasks on demolition projects, and such removals are governed by the EPA and various state regulations including some by state fire marshals. Inspection and removal as well as transportation and disposal of soils suspected to be contaminated by POLs are frequently performed by the demolition contractor or his subcontractor.

Biological Hazards for the most part consist of bird or bat droppings but can be any organic material with the potential to pose a danger to humans. The EPA regulations govern the removal and handling of such hazards.

Mold appears in hundreds of types, but there are only a few that are hazardous to health. Other than guidance from OSHA and recommendations on mold remediation from the EPA and state or local health departments, there are currently no regulations covering the removal or disposal of mold.

Radiological Hazards can occur when instruments containing radioactive isotopes have not been properly removed or when the demolition involves removal of construction features that have become contaminated by exposure to radiation. This may occur in parts of a nuclear reactor or a test area where radioactive sources previously existed.

Chemicals of many kinds can be present at demolition sites, particularly industrial sites. Regulations exist for handling many hundreds of chemicals from the common to the exotic.

Creosote was a popular chemical used for the protection of railroad ties and flooring blocks. These products are currently exempt from hazardous disposal requirements in most areas. In some instances, creosote oozing from the wood (bleeding) may be subjected to hazardous waste disposal procedures.

Dust is becoming more of a concern, especially in urban areas. For many years, the use of fire hoses with adjustable nozzles has been the method used to control dust. Today, there is an assortment of equipment that can project a fine water spray up to 200 feet to capture the dust particles before they leave the immediate worksite. Regulations exist that cover nuisance dust as well as dust with the potential to pose health hazards at the jobsite. Personal exposure to lead dust as well as silica, a frequent component of dust, especially during brick, masonry, plaster, or concrete demolition, is regulated as well.

Figure F.02. Dust control.

Mud—Most municipalities require that streets outside of demolition projects be kept clean. This is accomplished by either sweeping or washing on a regular basis. Trucks may also be required to pass through a wheel wash station prior to leaving the site.

Figure F.03. Wheel wash before leaving site.

Fluorocarbons—Because of their damaging effect on the earth's ozone layer, fluorocarbons such as Freon used in refrigeration/air conditioning systems along with Halon used in fire extinguishing systems must be removed without release to the atmosphere before demolition begins.

The examples of hazardous materials listed above cover the most typical, but not all of the hazardous materials that may be encountered on a demolition project.

In summary, the regulations that govern aspects of handling hazardous materials are available online and through printed materials provided by many regulatory agencies. It is important for the demolition contractor to be informed as to the nature and specific requirements for dealing with each of the hazardous materials that may be encountered on a project.

G. STATE AND LOCAL REGULATORY AUTHORITIES

The federal government is not the only organization that creates regulations to be followed when performing demolition operations. For example, the state of California requires that a special demolition permit be obtained from CalOSHA if a building to be wrecked is over thirty-five feet in height. Many states and municipalities have demolition regulations, some of which are more stringent than the federal regulations.

Examples of non-federal regulations include the following:

- Use of municipal off-duty police to act as flaggers
- Special requirements for the use of city fire hydrants
- Restrictive work hours in certain locations
- Requirements to use certain haul routes for trucking debris and moving equipment
- Haul restrictions—truck size, tarping requirements, weight requirements
- Requirements for special specifications for barricades and pedestrian walkways at demolition sites
- Local business licensing and permitting
- Restrictions on work methods such as no use of wrecking balls or implosions
- Restrictions on performing any work in a waterway due to fish migration

It is strongly recommended that anyone having responsibility for estimating or managing a demolition project be thoroughly familiar with the regulations that will affect the project. Failure to take regulations seriously can endanger lives, limit a demolition contractor's ability to obtain future work, or may even result in financial catastrophe.

H. STUDY QUESTIONS

1. True or False. The OSHA standards do not carry the force of federal law, and companies that are found to be violating the OSHA standards do not face jail terms.

2. What OSHA standard covers respiratory protection?

3. What ANSI standard reflects "Safety and Health Program Requirements for Demolition Operations"?

4. List the three "Environmental Site Requirements" in the OSHA/EPA standards.

5. What is the proper procedure if hazardous materials are found during demolition?

6. What is the current regulation covering removal and disposal of mold?

References

National Demolition Association. 2004. *Demolition Safety Manual*. Doylestown, PA: NDA.

National Demolition Association. 1999. *Hazard Communication Program*. Doylestown, PA: NDA.

CHAPTER 6

ESTIMATING—QUANTIFYING AND PRICING THE DEMOLITION PROJECT

A. INTRODUCTION

Estimating a demolition project can be far more challenging than estimating a typical construction project. Although there can be similarities between various demolition projects, no two projects are exactly alike. Estimating factors that can have a significant effect on the price of a demolition bid may include, but are not limited to, the following:

1. Preparatory Estimating Tasks
2. Regulatory Requirements and Restrictions
3. Project Location/Accessibility/Adjacent Buildings and Infrastructure
4. Project Size/Scope of Work
5. Construction Type and Condition
6. Available Information
7. Available Resources
8. Salvage
9. Schedule
10. Interfaces
11. Weather
12. Disposal Options/Haul Distances and Restrictions
13. Hazardous Materials and Other Contamination

In this chapter, factors common to all demolition projects will be examined followed by sample estimates of different projects types.

B. PREPARATORY ESTIMATING TASKS

Specifications for the work will normally be provided by the clients' architectural/engineering firm. It is imperative that the estimator thoroughly understand the specifications and lists those sections that will require cost estimates and/or information such as corporate experience, insurance, bonding, safety records, subcontractor list, work plans, and other required information.

It is also important to determine a Rough Order of Magnitude (ROM) cost for a project so that if the project is too small or too large to match available resources, the company can move on to more worthwhile opportunities.

B.1. Regulatory Requirements

The demolition estimator must be familiar with the regulatory requirements at the federal, state, county, city, and owner level that affect the cost of the work. Some of these regulations may also affect the means and methods to be used to perform the work. (See chapter 5 for a discussion of regulatory requirements common to demolition projects and a listing of some applicable regulations.)

At the federal government level there are several agencies that regulate construction and demolition activities. Listed below are the most common regulations and the agencies that publish them:

TABLE 6.01

Regulatory Agency	Governmental Department
OSHA (Occupational Safety and Health Act)	Department of Labor: CFR 1926
NPDES (National Pollution Discharge Elimination System)	Environmental Protection Agency (EPA): CFR Parts 9, 122, 123,124
NESHAP (National Emission Standards for Hazardous Air Pollutants)	Environmental Protection Agency (EPA); also known as the "Clean Air Act"
CERCLA (Comprehensive Environmental Response, Compensation, and Liability Act)	Environmental Protection Agency (EPA); also known as "Superfund"
RCRA (The Resource Conservation and Recovery Act)	Environmental Protection Agency (EPA): CFR Parts 260-265

There is a real cost to meeting the requirements of the myriad of regulations, and the estimator must be aware of these costs and include them in the estimate. For example, the environmental work required to comply with Storm Water Pollution Prevention Plan (SWPPP) for sites over one acre can cost thousands of dollars to install and maintain. In general, the regulations are beneficial and, when properly executed, provide for a safer and more environmentally responsible project.

Since regulations vary from area to area, the estimator must not assume a job in New York will have the same regulatory requirements as one in Arizona. Use of the Internet to check applicable regulations that will be required and the costs of compliance is advised.

It is important during the bidding process to inform potential subcontractors, such as an electrician responsible for re-routing power lines, that they must include the costs of regulatory requirements specific to their trades. Subcontractors must also include costs of complying with federal, state, and local regulations that affect all operations at a particular jobsite.

Demolition and other work related to a specific project usually require a permit(s) issued by the local municipality. In addition to a "Demolition Permit," permits are often required for Asbestos Containing Materials (ACM) removal, SWPPP, street use, extended work hours, excavation, utility work, Underground Storage Tank (UST) removal, paving, and various special permits. The estimator needs to be aware of the permits required for a particular project and their costs, which can exceed several thousand dollars.

B.2. Project Location/Accessibility/Adjacent Buildings and Infrastructure

It is much less expensive to perform the demolition of a multi-story building surrounded by parking lots than to demolish the same building in a congested urban setting. Another example of cost impact is demolition work in operating industrial plants. The demolition estimator must take the location factor into account when developing their cost estimates. The cost of protecting other structures and utilities while performing demolition work can account for a significant portion of the overall cost.

Most urban and industrial project sites provide adequate access to permit the use of typical demolition equipment. For some sites, accessibility can be difficult, and less efficient means must be utilized for the work. A common example is demolition within an operating facility that may require the use of small equipment and hand labor rather than the use of a large crane or excavator, which, in most instances, would be less costly.

Some sites are located in remote areas and require considerable planning and effective management to accomplish the mobilization/demobilization of equipment as well as the supply and housing of the labor force. Location factors, which can add considerable cost to a project are:

- Temporary living quarters, usually motels
- Personnel transportation

- Provisions for supplying fuel
- Security
- Transportation of specialized equipment and tools

B.3. Project Size

The size of a demolition project usually has a significant effect on the unit cost. For example, assume Building A is 3,000 square feet and was estimated to cost $8.00/sf, whereas Building B is 20,000 square feet and may cost only $5.00/sf. The unit cost difference is the result of costs being spread over a larger area. For example, mobilization and demobilization may cost $5,000 for either of the buildings, but would result in a unit price of $5,000/3,000sf or $1.67/sf for building A, yet would cost only $5,000/20,000sf or $0.25/sf for building B. Other factors, such as differences in the efficiency of the operation, local labor rates, and project location can contribute to unit cost differences.

B.4. Available Information

The accuracy of a demolition estimate is dependent upon the information available to the estimator and requires the estimator to attempt to obtain information from any of the sources listed below:

B.4.1. Demolition Company Records. Good records of previous work are an invaluable resource for the demolition estimator. Such records can provide a starting point and cross-checks for determining quantities and costs. For example, if a previous demolition project of a similar nature resulted in the average production of 300 square feet of debris per truckload, then it is likely the project being estimated would have a similar production.

B.4.2. As-Built Drawings of the Structures. Whenever possible, the estimator should try to obtain copies of the as-built drawings of the structures to be demolished. As-built drawings will assist him or her in accurately quantifying the building components, particularly the foundations and other structural features. The process of measuring and calculating the quantities of a structure is referred to as the "takeoff." It may be necessary to be persistent in order to obtain drawings, but they can save both time and increase the accuracy of the takeoff.

B.4.3. Information Regarding Utility Locations. Most drawings furnished to prospective bidders will contain whatever utility information that is available to the owner. However, this is not always the case. The estimator needs to assure himself or herself that sufficient costs are allowed to locate unknown subgrade utilities or stipulate in the proposal that costs for utility surveys are not included in the base bid.

B.4.4. Ability to Thoroughly Inspect the Entire Site. Failure to allow enough time to thoroughly inspect a demolition site is perhaps the primary cause of a poor estimate. The site visit is the best opportunity for the estimator to verify drawings and to physically determine the effects of added costs such as protecting nearby structures or the difficulty of working around high-voltage power lines. Estimating costs without seeing the site is to be avoided if at all possible.

B.4.5. Location of Acceptable Disposal and Recycling Sites. It is likely that the estimator will already be familiar with those businesses that provide landfill and recycling yards in their general work area. However, "going out-of-town" or away from familiar work areas requires that the estimator carefully investigate the reputable firms in these areas.

B.4.6. Former Uses of the Property. When a site is visited, what is seen may not be what it appears. Demolition project sites may have had previous "lives" as industrial operations, and there may have been previously demolished construction materials buried on the site. The possibility that petrochemical materials or other contaminants are present on a site can have a major cost impact and should be investigated before completing the bid.

B.4.7. ACM Survey. If there is one issue that is almost always a major cost factor in any demolition project, it is the removal of Asbestos Containing Material (ACM). The cost of asbestos abatement can sometimes exceed that of the demolition of the structure itself.

B.4.8. Hazardous Material Surveys (Haz/Mats). The estimator must be aware of the quantities and the locations of other hazardous materials (other than ACM) when estimating the costs of removing all haz/mats. Often, the demolition contractor will subcontract the removal of hazardous materials.

B.5. Available Resources

The estimator must be familiar with the capabilities of their company in order to develop a realistic estimate of the costs of a project.

B.5.1. Availability of Personnel. The demolition estimator should be aware of the qualified personnel available for a given project. It may be necessary to consult with management to determine whether it can be assumed that the appropriate personnel will become available should the company be awarded a particular job. For example, when estimating a project that requires working in a congested downtown area and the company does not have available supervision experienced in that type of work, the estimator should question management as to whether the company should bid that job. If so, the estimator should make cost allowances for bringing the proper personnel on board for the project.

B.5.2. Equipment Availability. The estimator usually assumes the equipment needed for the job will be available from his or her company, or can be readily rented. If the company does not have the equipment necessary to perform a particular type of project during the scheduled time, the estimator must factor the cost of renting the necessary equipment into the bid.

B.6. Schedule

The estimator will, at some time during his career, encounter projects that require unrealistic performance schedules. As a practical matter, this means that sufficient costs must be added to account for late completion penalties, labor overtime costs, as well as additional personnel and equipment that will be required to meet the demolition schedule. An unrealistic schedule may be cause to choose not to bid a job.

B.7. Salvage

Unique to the demolition industry is the effect of salvage on the cost of the work. In order to produce a competitive estimate, the demolition estimator must take into account the net values of equipment and materials that will be removed during the process of demolition or dismantlement. Note: The estimator must be aware of any items that the owner/client may wish to keep. Misunderstanding the ownership of items having potential salvage credit can create severe financial and legal problems.

Typical salvage materials generated during the demolition of a structure, particularly industrial facilities, may include the following:

- *Steel and Iron Scrap*—structural steel, pipe (both steel and stainless steel), reinforcing steel bars, and various types of sheet metal
- *Non-Ferrous Metals*—copper wire and buss bar, aluminum wire, alloys, conduit, fixtures, and brass fixtures
- *Timber*—any type of sound timber, free of rot, and at least four by six inches in nominal dimension, lumber and rare, exotic woods
- *Architectural Features*—this category can include a wide variety of saleable decorative items such as doors, stone, and terra-cotta items
- *Brick*—the value of various types of brick varies greatly from region to region as does the cost of cleaning the brick for sale

- *Electrical Equipment*—modern electrical panels, switch gear, transformers, motors, generators, and associated equipment
- *Process Equipment*—industrial demolition projects may contain useable process equipment such as overhead cranes, boilers, tanks, compressors, heat exchangers, conveyors, and a wide variety of items. Some equipment may be valuable for parts only.
- *Obsolete Equipment*—may be valuable as "scrap only" or may be saleable for re-use in third world countries

B.8. Interfaces

Most demolition projects will have one or more interfaces with other activities on the site. The estimator must be aware of such circumstances and allow costs for specified as well as consequential interfaces. Typical interfaces include but are not limited to the following:

B.8.1. Other Contractors. The demolition contractor is often a subcontractor to a general contractor who controls the site and the overall project. Other subcontractors may also be working in the same area as the demolition contractor. Therefore, the estimator must consider the resulting delays and/or restrictions in their plans for the work such as limitations on equipment and production.

B.8.2. Utility Work. Terminating and/or re-routing of utilities, particularly for industrial projects, often require the demolition contractor to temporarily stop work or to work in a manner that inhibits normal production. The estimator must take the costs that are associated with the utility interfaces into account.

B.8.3. New Construction. The effect of new construction on a demolition project can add extra costs to the demolition estimate. The estimator must be aware of any restrictions imposed by construction activities such as not making exhaust smoke. The cost impacts of such restrictions may result in schedule delays and requirements for overtime work that add costs to the estimate.

B.9. Weather

The time of year and unusual weather events can result in additional costs for a demolition project. The cost impact of weather is very difficult to assess. If the work must be done in an area of frequent extreme weather events, the demolition estimator should take this fact into account and allow some contingency for the influence of weather in the cost estimate. As a standard practice, some firms add a contingency allowance to account for weather-related costs.

C. THE DEMOLITION ESTIMATE

C.1. Quantity Takeoff

The estimator for any demolition activity must do his or her best to develop a reasonable estimate of the quantities that are to be demolished. (Please refer to the Appendix for a sample "Estimator's Pre-Bid Checklist"). Without such an assessment of the quantities, the cost estimate becomes a guess and can be either unrealistically high or low. Regardless of what information is available to the estimator, nothing is more important than a visit to the site and an opportunity to inspect the project firsthand. Such a visit will give the estimator an opportunity to verify drawings, measure building materials, note any complications in performing the work such as location of utilities, proximity of other construction, ground conditions, and any other information that will affect the cost and schedule for performing the work.

The estimator should use all the resources available to determine the following project quantities:

- *Hazardous Materials (Haz/Mat)* such as ACM, Polychlorinated Biphenyls (PCBs), mercury containing items such as lighting and switches, petroleum products, and a variety of other materials.

These quantities are sometimes available from professionally produced surveys that are part of the project specifications. If they are not available, the estimator should either request this information or stipulate in his proposal to the client that this information was not available and the cost proposal excludes the costs for doing this Haz/Mat removal work. If quantities of hazardous materials exceed those provided in a survey used to bid the project, the demolition contractor should seek additional compensation.

- *Building Materials* such as concrete, brick, concrete masonry units/concrete block (which are commonly referred to as CMUs), wood, miscellaneous debris, steel and other metals, backfill and site restoration, as well as whatever other tasks may be required by the specifications. These quantities are seldom furnished to the estimator and must be calculated from available resources such as drawings and site measurements.

- *Site Restoration* such as size and type of backfill material, compaction, landscaping, fence construction, and anything else required by the specifications after the demolition itself has been completed.

As-built drawings are the most useful source for measuring quantities; however, they are often not available for most demolition projects. Even if drawings are available, the estimator needs to carefully inspect the site and assure himself or herself that the drawings are reasonably complete and accurate.

In addition, the estimator must determine those quantities that fall into the negative cost column. There may be quantities of salvageable materials that can drastically reduce the net cost of the work. Drawings can be very helpful to the estimator in quantifying materials that have salvage value.

Mistakes can be made in every human endeavor. Demolition takeoff estimates are no exception. The estimator should always have a fellow estimator or management personnel review the takeoff. Sample forms for tabulating demolition quantities are included in this chapter.

C.2. Production Estimates and Unit Costs

In order to produce a realistic cost estimate for a demolition project, the estimator must have knowledge of the production rate to be expected for each of the tasks to be performed. As an example, assume that one of the tasks in a demolition project might be the wrecking of a wood frame, two-story apartment building. Assume that from prior experience, the estimator knows that a thirty-five ton excavator can wreck about 5,000 square feet of such a building in an eight hour shift. With that information, the estimator can calculate the total cost of labor, support equipment, and the excavator. Dividing the total cost by the production estimate for that task yields the cost per square foot. Table 6.02 is an example of a unit cost calculation.

Table 6.02. Demolish a two-story apartment building at a rate of 5,000 square feet per day.

EQUIPMENT	No. Each	Cost/Day Equipment	LABOR	Each	Cost/Day Labor
Excavator-35 ton	1	1,000	Oper. Engr.	1	360
Skid Steer Loader-1/4cy	1	264	Laborer	1	280
Water Truck	1	280	Laborer	1	280
Sub Totals		1,544			920
Base Cost per Day (Equip + Labor)				>>	$ 2,464
Base Unit Cost		equals	$0.49 per sf		

(*Note: This sample represents only the basic cost of demolishing the building and includes no allowance for removal, disposal, or any other costs.*)

Many unit costs may be used when estimating a demolition project. For example, there might be square or cubic, foot or yard, pound or ton, and load or mile unit costs for other tasks such as:

- Concrete superstructures
- Concrete slabs-on-grade

- Concrete foundations
- Concrete crushing
- Wood frame buildings
- Composite assemblies such as combinations of various materials
- Steel frame buildings and structures
- Costs of handling, trucking, and disposal of any of the typical building components

Unit costs for demolition are based on rates of production depending on the equipment and labor to be used. Accurate records of previous labor and equipment used, and the production achieved, provide the information needed to develop reasonably accurate unit costs. Average labor rates and equipment rental rates may also be found in the *R. S. Means Facilities Construction Cost Data* reference book published annually by Reed Construction Data.

Refer to Table 6.07 for a sample form that can be used on a computer spreadsheet or similar format to develop unit costs for any typical demolition task.

C.3. Base Costs

This term is used to capture those costs that are a part of the work to be accomplished. They include everything from the cost of submittals and permits to performing all the functions of the demolition project.

C.4. Overhead Costs

These are the costs that cover the expenses required to be in business, regardless of whether or not any work is being accomplished. Examples of overhead include costs for operations of the home office such as rent or mortgage, insurance, legal support, marketing, office and shop utilities, and any other costs not directly associated with a particular project. The costs of owning and maintaining equipment are not normally covered in the overhead costs. Overhead costs are typically in the 10% to 20% range of the gross sales of the company. Refer to chapter 7, "Contracts and Accounting for the Demolition Project," for additional discussion of these costs.

C.5. Salvage

A major component of any demolition project is the estimate of the value of salvage, which is expected to produce income. The net value of salvage is computed by deducting the cost of salvage operations from the value of the salvage. This amount is usually referred to as "salvage credit." In some circumstances, the recycled value of metals from an industrial project can actually exceed the cost of the demolition.

To be competitive, the estimator must have a reasonably accurate takeoff of those materials that can be salvaged and recycled as well as the unit value of the salvaged materials at current market prices. See section B.7 above for further discussion of salvage.

C.6. Bid Price

This term is often referred to as the "net cost." It should be noted that this is the "net cost" to the owner, not the contractor, because it includes profit. It is calculated by factoring in the following components: Bid Price = Cost of the Work + Overhead Costs + Profit – Salvage Credit. Each of these components is calculated by adding together several sub-components that will be shown in detail in the sample estimates later in this chapter.

C.7. Final Review

Most companies will conduct a final review of a bid before submitting the completed bid documents to the client. It is helpful to have more than one individual review the bid for completeness and verify that there are no mathematical errors or confusing language. Any review should include the following tasks:

a. Documentation

- Check to be sure that the bid includes all the requirements of the specifications
- Required Insurance Certificates with proper coverages, limits, named insureds
- Check that the Bid Bond is in the correct form and complies with the specifications
- Check to be sure that any subcontractors are listed as specified
- Check to be sure that any forms requiring information for utilization of minority, disadvantaged, or other categories of businesses have been properly completed
- Check to be sure that requirements for signatures and seals have been properly executed
- Check to be sure that any requirements for safety information is included
- Check to be sure that any requirements for work plans, that is, demolition, environmental, schedules, etc., have been completed and included
- Verify that arrangements have been made to deliver the bid on time

b. Pricing

- Check quantity summaries and verify they are reasonable for the work required
- Calculate unit prices where appropriate, and verify they are reasonable for the project and are in line with previous company experience
- Check to be sure that all pricing requirements are completed and are accurately tabulated

D. SAMPLE ESTIMATE FORMATS

Before reviewing samples of typical estimates, it is necessary to organize the information that will be used to create the estimate. The examples shown below are intended as guides for collecting and presenting data in logical formats that collectively will result in a demolition estimate. Electronic versions of some of these spreadsheets are available at http://docs.lib.purdue.edu/demolition/.

D.1. Quantities

To reduce the possibility of serious errors, the raw data, whether taken from drawings, field measurements, or provided as part of the specifications, must be organized in such a manner as to be easily understood. For this discussion, it is assumed that hazardous materials information will be provided by a Survey Report prepared by accredited specialists.

D. 1.1. Concrete, Concrete Masonry Units (CMUs), Masonry. This category of materials will include construction features such as foundations, slabs, walls, above grade floors, and roofs. Table 6.03 below is a sample format of a takeoff summary for a typical concrete structure. Depending on the circumstances, more detail can be added. In this example, no allowance was made for the deduction of a portion of the exterior walls for windows and doors. This and other quantity additions and deductions can easily be factored into the calculations as needed.

D. 1.2. Wood (timbers, lumber, sheathing, roofing, and miscellaneous materials). This class of materials will include construction features such as entire buildings, floors, walls, ceilings, partitions, roofs, and composite construction like fiberglass panels on wood studs.

In Table 6.04, wood and miscellaneous materials present in the sample estimate are calculated.

D. 1.3. Steel (structural, corrugated iron, pipe, machinery, reinforcing bar, and miscellaneous items). Steel may be found everywhere on demolition projects, either as the primary building support system or for parts of buildings including such items as canopies and mezzanines. Piping is also found in most buildings along with stairs, window casings, air handling ductwork, and a wide variety of other steel products. There may also be other types of metals in sufficient quantities to justify a quantity estimate.

D. 1.4. Earthwork and Site Restoration. This category of work can include a wide variety of tasks required to produce a finished project from a grassy field to an asphalt parking lot. Shown in Table 6.06 is the quantification of typical site restoration items.

Table 6.03. Concrete Example.

CONCRETE, CMU, BRICK, TILE - Estimated Quantities							Date:		5/26/08	SAMPLE ONLY		
Item No.	Description	L per ft	W ft	Area/ sf	h ft	t ft	Conc Fndn cy	Conc SOG cy	Conc Struct'l cy	Conc Total cy	CMU/Brk cy	Fill cy
>>	this project is the demolition of a 3 sty concrete apartment building with basement											
	Footprint	100	40	4,000								
	Stories			3								
	Total Area			**12,000**								
	basement slab			4,000		0.5		74				
	basment fndn	280			1.5	2.0	31					
	basement wall	280			9	0.7	65					
	floors 1,2 & 3			12,000		0.8			356			
	roof			4,000		0.8			119			
	exterior walls	280			30	0.7			218			
	int walls, stair wells, etc	40			39	0.7			40			
	Total Concrete			**12,000**			**96**	**74**	**732**	**903**		
	Basement fill -add 30% for compacti			4,000	10							**1926**

Table 6.04. Wood Example.

WOOD - Estimated Quantities (SAMPLE ONLY)				Date:			5/26/08		
Item No.	Description	L per* ft	W ft	Area/ sf	h ft	bf/ sf	wood debris mbm		
>>	this project is the demolition of a 3 sty concrete apartment building with basement								
	Footprint	100	40	4,000					
	Stories			3					
	Total Area			**12,000**					
	basement partitions	100	9	900		2.5	2.3		
	3 floors of partitions-same	200	24	4,800		3	14.4		
	ceilings flrs 1,2 & 3			12,000		1	12.0		
	ext wall furring	280	24	6,720		1.2	8.1		
	miscellaneous junk allowance						2.0		
	roofing - built-up			4,000		1	4.0		
	Total mbm						**42.8**		
	assume that drop boxes will be used to haul the wood debris to the disposal site:								
	allow 3 mbm in each box, therefore 42.8mbm / 3 mbm = 14.3 boxes required.								
	mbm = 1,000's of board foot measure: 12 bf = 1 cubic foot								
	per* = equal perimeter measurement								

Upon completion of the demolition takeoff and summarization of other quantities, the estimator must prepare a draft copy of the actual pricing for the job. This task can take many forms, and there are dozens of formats that are used by demolition contractors. For this discussion, it is assumed that the estimator must develop unit pricing for the demolition and removal of a concrete and wood structure with an attached steel frame storage area. The quantities estimated for this example are described in the four takeoff summaries shown in Table 6.03 through Table 6.06. Table 6.07 is an example of a unit price

Table 6.05. Steel Example.

STEEL - Estimated Quantities								**Date:**		**5/26/08**	
(SAMPLE ONLY)											
Description	**L**	**W**	**Area**	**unit**	**Wt per**	**No.**	**Total**	**Sheet**	**Pipe**	**Scrap**	**Total**
	ft	**ft**	**sf**	**wt**	**each**	**Pcs**	**Wt**	**metal**	**Wt**	**Mach'ry**	**Scrap**
				lbs/lf	**lbs**	**ea**	**ton**	**wt**	**ton**	**ton**	ton
this project is the demolition of a 1 story steel frame shop with corr. Iron roofing & siding											
Footprint	100	40	4,000								
Stories						1					
Total Area			**4,000**								
columns - wf 8"x10"	20			39	780	14	5.46				
girts - channel 8"	20			11.5	230	56	6.44				
trusses	40			60	2,400	6	7.2				
purlins-wf 8"x6"	11			24	264	105	13.86				
bracing L-3"x3"	28			7.2	202	12	1.21				
crane rails	20			80	1,600	10	8				
overhead crane est					0		0			15	
siding	280	20	5,600	2		1		5.6			
roofing			4,000	2		1		4.0			
misc machinery-est.										5	
pipe 4"	200			10	2,000	1			1		
Total Scrap Wt							**42**	**10**	**1**	**20**	**73**
Unit Weight										36.4	lbs/sf

Table 6.06. Site Work Example.

SITE RESTORATION - Estimated Quantities								**Date:**		**5/26/08**
	(SAMPLE ONLY)									
Item	**Description**	**L**	**W**	**Area**	**Depth**	**Vol**	**No**	**Total**	**Grading**	**Seeding**
No.		**ft**	**ft**	**sf**	**ft**	**cy**	**each**	**Vol**	**Area**	**Area**
								cy		
>>	this project is the site restoration of an area once occupied by a warehouse									
	Footprint	200	100	20,000						
	excavate for footings	6	6	36	3	4	100	400		
	backfill basement	100	50	5,000	10	1,852	1	1852		
	import backfill x 1.3*							2928		
	grade area-allow 20% more			20000					24000	
	seeding									24000
	Total Import							2928		
*	(allow for 30% compaction)									

spreadsheet for the calculation of the demolition of the structure based on the volume of concrete. Note that the unit cost shown is for demolition only without any other tasks, such as hauling and disposal, or overhead. Similar calculations will need to be made for each of the tasks that are listed in Table 6.08, "Sample Demolition Cost Estimate Summary." Many of the unit prices will be available from company personnel or records of previous jobs.

Table 6.07. Unit Price Calculation Example.

UNIT PRICE CALCULATIONS - 2008				Category	Concrete Bldgs & Structures Above Grade		
Project:	Estimate:Fig. 3.C.01 - Sampl	Date:		29-May-08		Lab Rates	$/hr
Task:	Demolish and stockpile concrete structures				Unit	Op Engr	42
Location:	all	Quantity Base:		733	cy	Laborer	27
Task:	>>>>>> assume	12,000	sf				
		Total A	12,000	SF			

Crew:	Op Engr	Laborer	Exc/UP	Exc/B&T	Loader	Skid	Shear/UP	Man Lift	Truck	Water
No.	2	4	100k#class	70k#cla	3 Cy	Steer				Truck
Rate/hr	84	108	212	154		33	50			35

Cost/hr >	$676	Cost per Day >	$5,408	Total Days >	3.665

Production	Unit	per hour	per shift
	cy	25	200

Unit Cost	Labor	Equip	Total	
per Cy	$7.57	$19.47	$27.04	per cy

Percent:	Labor:	28%	Equip:	72%	Other:	0%

Add Mat'ls	Item	Cost
	Total	$0

Special Add-ons	Item	Cost
	Total	$0

Unit Cost Add-Ons:		Mat'ls	Special	Total
		$0	$0	$0

TOTAL UNIT COST:	$27.04	per CY
Alternate Unit Cost:	$1.65	per SF

The following "Sample Demolition Cost Estimate Summary" is a system for organizing the typical costs for a demolition project. Such a format is useful as a checklist so that the estimator can verify that all of the typical costs that are associated with demolition are accounted for in the Cost Summary. The Sample Quantity Calculations are a hypothetical quantity summary of the primary construction materials of a typical, simple structure.

Sample Quantity Calculations

The following assumptions are used for the quantities estimate:

- Structure is a three-story apartment building measuring 100 x 40 feet, or 12,000 square feet
- Structure has a full basement of 4,000 square feet
    - basement walls are 0.67 cf/sf
    - basement slab is 0.5cf/sf
    - perimeter footings are 3 feet wide and 2 feet deep
- Exterior walls, floors, and roof are poured-in-place concrete construction
    - floors and roof average 0.75cf/sf
    - exterior walls average 0.70 cf/sf
    - partitions are 0.17 cf/sf

- Interior partitions and ceilings are drywall construction
- Contractor shall construct a temporary security fence
- Contractor shall cap utilities at the property lines
- Contractor will pay $1,500 for permits
- Mobilization and demobilization of equipment is calculated to be $7,500
- Primary piece of equipment will be a fifty ton class excavator
- Backfill is available on-site at no cost

Table 6.08. Sample Demolition Cost Estimate Summary.

SAMPLE DEMOLITION COST SUMMARY SHEET	**Date:** **Project:**		**Gross Area - sf**		**12,000 sf**
Item	**Activity Description**	**Units**	**Quant**	**Unit**	**Basic**
				Cost	**Cost**
	Demolition				
1	Permits	ls	1	1500	$ 1,500
2	Mobilization/Demob	ls	1	7500	7,500
3	Engineering	hrs	0		0
4	Temp Fence	lf	440	6	2,640
5	Site Protection	ls	1	1500	1,500
6	Misc Supplies/Mat'ls	ls	1	500	500
7	Utility Disconnects	ea	3	500	1,500
8	Salvage Work	ls			0
9	Sever Structure	ls			0
10	Gut Interior	sf	12000	0.50	6,000
11		ls			0
12	Demo Structure Above Grade	cy	703	25	17,575
13	Demo Foundations	cy	94	35	3,290
14	Demo Slabs	cy	74	9	666
15	Demo AC & Misc Conc/Masonry	cy			0
16	Load & Haul Wood Debris	ld	9	346	3,114
17	Prepare Concrete for Crushing	ton	1594	5	7,970
18	Crush Concrete—stockpile on site		1740	6	10,452
19	Load Scrap— (use buyer's drop box)	ton	48	11	528
22					0
23					0
24					0
	Subtotal Demolition:				**$64,735**

Item	Activity Description	Units	Quant	Unit Cost	Basic Cost
	Site Restoration				
25	Excavate	cy			0
26	Backfill—including compaction	cy	1480	6.5	$9,620
27	Grading—(25% greater than footprint)	sf	5000	0.15	750
28	Seeding—for small area	sf	5000	0.25	1,250
29					0
30					0
	Subtotal Site Restoration:				**$11,620**
	Project Support				
31	Supervision Allowance	days	8	600	$4,800
32	Per Diem	md	24	85	2,040
33	Travel	ea	5	250	1,250
34	Job Office and Temp Utilities	days	8	200	1,600
	Subtotal Project Support:				**$9,690**
	TOTAL DEMOLITION COST:				**$86,045**
35	Disposal of Debris	ton	126	50	$6,300
36					
	TOTAL DISPOSAL COST:				**$6,300**
37	**Environmental Costs**				
EXAMPLE ONLY	Allowance Only From Other Worksheets				**$10,000**
	TOTAL ENVIRONMENTAL COST:				**$10,000**
	TOTAL BASE COST:				**$102,345**
38	**Add for Overhead Cost**			15%	**$15,352**
39	**Add for Profit**			10%	**$10,235**
	Sub Total				**$127,932**
40	**Less Credit For Salvage**	ton	48	(150)	**($7,200)**
	TOTAL BID PRICE				**$120,732**

No two demolition contractors will use the same format, however, in one form or another, the examples illustrated herein are straightforward and simple to use. Of course, in a real estimate, these forms would be set up with a computerized spreadsheet.

E. COMMON ERRORS AND PITFALLS

E.1. Preparing the Estimate—Errors and Pitfalls

- Incomplete understanding of specification requirements
- Failure to allow sufficient time for site inspection and therefore not having time to thoroughly inspect site
- Failure to verify structure dimensions, size, and type of construction materials
- Using incorrect scales with drawings
- Failure to contact potential subcontractors in time for them to produce competitive bids
- Guessing the cost of unfamiliar subcontract and/or material costs
- Failure to verify salvage quantities and values

E.2. Administrative Functions—Errors and Pitfalls

- Failure to order Bid Bonds in a timely manner
- Failure to complete all information required by Bid Documents
  - subcontractor list
  - disadvantaged business enterprises contacts and listings
  - listing of requested licensing
  - verify company has proper insurance in force
- Failure to acknowledge addenda

F. STUDY QUESTIONS

1. Make a list of the information that is needed to be able to create a bid for demolishing a small bridge over a small stream using the following conditions:

- No demolition materials will be allowed to fall into the water
- There is adequate access to the bridge

2. Using the Internet, determine the current prices being paid in Chicago for the following scrap metals:

- #1 steel scrap
- #2 steel scrap
- #1 copper scrap
- #1 aluminum scrap

3. Assume a potential client is on the phone and asking if your firm would be interested in a project they are planning. Create a list of questions you would ask in order to determine if this project is of interest to your firm.

4. Using the following conditions, calculate which landfill is most competitive and the difference in cost per load between the two landfills? Assume that there are two choices for disposal in landfills:

- Choice 1 is fifteen miles from the job and the tipping fee or landfill disposal cost fee is $25/ton.
- Choice 2 is five miles from job and the tipping fee is $60/ton.
- The trucks are hauling an average of fourteen tons per load and travel an average speed of forty miles per hour.
- Each truck will spend ten minutes at either landfill dumping the load.
- Loading time is fifteen minutes per load and the cost of truck and driver is $110/hour.

5. An estimator or management may have several projects bidding in the same general time frame. Create a list of factors that an estimator should consider before making a recommendation to management as to which project(s) make the most sense to bid.

6. Conduct an Internet search for information on Storm Water Pollution Prevention Plans (SWPPP) and assume your demolition site is on a fifty acre site located in EPA Region 8. List those items that must be addressed in accordance with the requirements of the SWPPP regulation.

7. Assume there are two locations for a six story concrete building.

- Location A is located in a downtown area with city streets on all four sides of the building.
- Location B is located at an industrial site with no obstructions within 100 feet of the building.

List those items that are likely to add to the cost of demolishing the building at Location A as compared to the building at Location B.

APPENDIX. SAMPLE ESTIMATOR'S PRE-BID CHECKLIST

- ☐ Is project appropriate?
- ☐ Determine Rough Order of Magnitude (ROM) of costs.
- ☐ Determine approximate time required to prepare bid.
- ☐ Order Bid Bond or Bid Deposit and Insurance Certificates.
- ☐ Make plans to visit site allowing sufficient time to thoroughly inspect the project.
- ☐ Determine if subcontractors will be needed.
- ☐ Contact appropriate subcontractors.
- ☐ Complete quantity takeoff.
- ☐ Contact potential salvage buyers.
- ☐ Complete draft estimate in a timely manner.
- ☐ Review estimate with management.
- ☐ Verify all parts of bid documents are completed and signed.

CHAPTER 7

CONTRACTS AND ACCOUNTING FOR THE DEMOLITION PROJECT

A. INTRODUCTION

The demolition business, as in any other business, requires proper documentation to remain viable. The wording of contracts and change orders to the contract can mean the difference between profit and loss on a project. It is not uncommon for a general contractor or an owner to use a one-sided contract format with prime contractors or subcontractors. It is up to the demolition contractor to review the contracts carefully, making changes and exceptions where necessary. It is also advisable to have all contracts reviewed by legal counsel.

B. LUMP SUM CONTRACTS

The most common form of contract is the "Lump Sum Contract," which means that the contract provides for a total fixed price to perform the work. Lump Sum Contracts are used either for contracts between owner and contractor or between contractor and subcontractor. The Lump Sum Contract format can be structured as a single price for all of the work or may be separated into phases that are awarded or withheld at the option of the owner or general contractor. An example of such a Lump Sum Contract might be as follows:

Phase 1: Demolish building down to top of slab

Phase 2: Demolish foundations and backfill voids

Changes in the work required are handled by "Change Orders," which modify the contract to the extent of the agreed price of the change order and the work to be done. Most change orders are additional prices for extra work, however, it is also possible to use deductive change orders, such as removing a portion of slab demolition from the contract to reduce the price of the work. Listed below are topics that are typically included in construction (demolition) contracts whether they are prime contracts or subcontracts:

- Names, addresses, and license numbers
- Contract amount and payment terms, including retentions and liquidated damages
- Bond requirements
- Insurance requirements
- Scheduling requirements
- Security requirements
- Requirements to comply with specifications
- Conditions for settling disputes
- Identification as to which state laws will be followed
- Conditions required for termination of the contract by either party
- Indemnification clauses
- Requirements for signatures by authorized parties

C. TIME AND MATERIAL CONTRACTS

A Time and Material (T & M) contract is an agreement between owner and contractor or between prime contractor and subcontractor whereby the cost of services to be provided is indefinite and the final price is indefinite. The contractor doing the work is paid on the basis of mutually agreed upon equipment and labor costs plus reimbursement for certain fees that might include permits, disposal fees, travel, per diem, miscellaneous purchases, overhead, and profit. These rates and prices become part of the contract. The actual language of the contract can be very similar to a Lump Sum Contract except for the provisions for furnishing T & M invoices.

The reason for entering into a T & M contract is usually because the scope of work cannot be accurately identified, and neither the owner nor prime contractor wants to take the risk of unknown work and unknown final costs. Examples of situations whereby a T & M contract might be considered the best approach include the following:

- Emergency demolition work of any kind where the scope of work and time required for a lump sum contract are not practical.
- Reducing the risks of subsurface demolitions, such as removing large concrete foundations and piling that cannot be inspected and quantified.
- An owner may want to perform some demolition to prepare for new construction and reduce future scheduling delays, but plans for the new construction have not yet been completed.
- An owner and a demolition contractor have a long-standing mutual trust for each other, and the owner is confident that the demolition contractor will provide the best job at fair prices.

For normal demolition projects, T & M contracts are not common. Generally, the owner wants to know what the "bottom line" is for project costs, and unless the T & M contract contains a "not-to-exceed" cost provision, there is the possibility that costs could get out of control.

From the demolition contractor's standpoint, a T & M contract can be a vehicle for undertaking risky projects without risking financial loss. For example, assume that an owner wants to build on a site formerly occupied by a chemical plant. Using the equipment and labor costs appropriate for this project, the demolition contractor can proceed under the direction of the owner's representative, and the T & M contract is the assurance that the contractor will get paid for all services. For example, if foundations thought to have been previously removed are discovered, and some of the soil is found to be contaminated, the demolition contractor can complete the job and does not have to worry about incurring any additional costs.

The disadvantage of a T & M contract from the contractor's point of view is that profits are limited. If the contractor develops a more efficient plan for the work, the owner is the only one who will benefit. In a Lump Sum Contract, the contractor would benefit from any innovations that would reduce costs.

D. GENERAL AND ADMINISTRATIVE COSTS

General and Administrative (G & A) costs or "overhead costs" are usually not directly chargeable to a particular project and may be referred to as the cost of doing business. These costs are incurred every month whether or not the contractor has any work. Although such costs will vary widely among demolition contractors, there are many G & A costs that are applicable to all contractors. The following list is an example of typical items covered by G & A costs:

- Salaries of corporate officers
- Salaries of office support staff
- Salaries of general superintendent, safety officer, and other professionals

- Salaries of full time shop and yard personnel (may also be charged to equipment ownership cost)
- Fringe benefits for salaried personnel including medical and vacation/holiday pay
- Retirement benefit costs for salaried personnel
- Outside accounting services
- Legal services
- Insurance—General Liability and Property Damage
- Cost of annual business licenses
- Advertising
- Cost of leases or amortization of real property improvements and buildings
- Utilities
- Office supplies and miscellaneous services
- Property taxes

Certainly not all of these costs will apply to every company, and there may be other costs not listed that will apply to a given firm. In order for the demolition contractor to recoup the costs for G & A expense, the person must assign a percentage cost to each project bid so that the G & A expenses are covered for a given year. Assume, for example, that demolition Contractor A has an annual G & A of $500,000 and his average gross sales are in the range of $3,300,000. The G & A rate for this contractor would be calculated as $500,000/$3,300,000 = 15%. A 15% charge to each project would cover contractor A's annual fixed G & A cost. Conversely, if Contractor B also has a G & A of $500,000, but has gross sales of only $2,500,000, his percent G & A would be $500,000/$2,500,000 = 20%. This would put him at a 5% disadvantage when bidding against Contractor A. Of course, there are many other factors that influence the outcome of a bid, but a lower G & A certainly helps in the competitive arena.

E. PROJECT OVERHEAD

"Project overhead" may be defined as costs that are directly chargeable to a particular project and may be included either as a group or simply as bid cost items along with the various production costs. The following list represents items that the demolition contractor may classify as project overhead costs:

- Project site office—office trailer and equipment
- Special insurance—for example, "Accidental Pollution" insurance
- Temporary utilities—power, toilets, etc.
- Per diem costs—often required in union agreements for out-of-area workers
- Travel for personnel
- Security—fencing and guard service
- Engineering services for approval of demolition work plans, etc.

The classification and estimating system that collects and accounts for project overhead costs will vary greatly among demolition contractors, but the costs are real and must be accounted for when developing cost estimates.

F. DOCUMENTATION

In the litigious environment of today, it is important for contractors to create and maintain accurate documentation of all phases of the work. For the demolition contractor, documentation may include items required by law as well as those items that are important to the efficient management of a business.

Documents that may be required include the following:

- Payroll records
- Employment records
- Income tax records
- Safety records, that is, OSHA 300 Log of accidents and hours worked
- Employee health records
- Engineering survey reports—required by OSHA for every demolition project
- Various records including the removal, transportation and disposal of hazardous materials, including ACM, PCBs, heavy metals, etc.

Each demolition contractor will have his or her own system for documenting operations and production. A daily report or project diary of all activities on the job is extremely important. It becomes a record of what was accomplished each day and other pertinent information. Daily reports can be of critical importance should there be a dispute between the owner and contractor at some later date.

A typical daily report should, at a minimum, include the following information:

- Names of all employees, their crafts, and hours worked
- Listing of all equipment on the job and hours worked
- Notations regarding any safety issues that were discussed and steps taken to mitigate any problems and improve safety
- Notations pertaining to any inspections by outside agencies, such as fire departments or OSHA
- Brief description of what was accomplished that day
- Notations regarding type, quantity, and weight of all items hauled off-site
- Weather
- Any accidents, incidents, etc.
- Photographic documentation of such things as accidents, work progress, unexpected site conditions, etc.

Production information contained in daily reports is beneficial to the demolition contractor for estimating future projects of a similar nature and may contain the following:

- A record of quantities and weights of various types of materials that were demolished and hauled or disposed of
- A record of any other pertinent production statistics

G. STUDY QUESTIONS

1. In what two cases would a lump sum contract most likely be used?

2. For emergency work of any kind where the scope of work and time required are not defined, what is considered the best contract approach?

3. What costs are identified as not directly chargeable to a particular project and may be referred to as the general costs of doing business?

4. When estimating a project and grouping costs, what general category would the following costs fall under?

- Temporary utilities—power, toilets, etc.
- Engineering services—for approval of demolition work plans, etc.
- Special insurance—for example, "Accidental Pollution" insurance.

CHAPTER 8

SAFETY ON A DEMOLITION PROJECT

A. INTRODUCTION

Entire books are dedicated to the safe operations of construction-related activities. The focus of this chapter is the general requirements for maintaining a safe workplace for the employees and general public that may be impacted by a demolition project.

Nothing is more important to a successful demolition contractor than a good safety program. Those companies that develop and maintain a culture of safe operations are usually the same companies that grow and succeed. There are a number of real benefits that are gained by the demolition contractor when he or she promotes and maintains a good safety program including:

- Protecting employee health and safety promotes a better, more loyal workforce
- Safety measures for the protection of the public and property promotes a good public image
- Insurance rates are lower for contractors with a good safety record
- Clients, both private and public, are more likely to seek proposals from contractors with a record of safe operations
- Low incidents of accidents reduce delays to the project

In general, a demolition contractor with a good safety record is likely to be running all facets of his or her business in a responsible manner and has a better chance of being profitable.

B. THE ROLE OF OSHA IN THE DEMOLITION INDUSTRY

In 1970, the U.S. Congress passed the Occupational Health and Safety Act and created the Occupational Safety and Health Administration (OSHA), which operates as an agency of the Department of Labor. OSHA, as it applies to the construction and demolition industries, is covered by the Code of Federal Regulation, CFR 1926. Demolition-specific standards are set forth in CFR 1926.850, Subpart T, "Demolition." The stated mission of the OSHA standards is to prevent work-related injuries, illnesses, and deaths by issuing and enforcing workplace safety and health.

When the law was originally enacted, each state had the choice of having their program administered by the federal government or the state could administer their own program as long as it was at least as strict as the federal program. Over twenty states now administer their own OSHA programs. It is important to note that when considering the implications and requirements of OSHA, the demolition contractor should make sure he or she is complying with the standard for each specific state.

Some of the regulatory topics particularly important for the safe working environment of a demolition project are listed below:

- Integrating safety into the job
- Worker training
- Engineering survey for structures to be demolished
- Hazard communication standard compliance
- Medical services and first aid
- Fire prevention and protection
- Site conditions and accessibility, such as lighting, ladders, scaffolding, personnel work platforms, and aerial lifts

- Public protection, such as protective structures, shoring, and structural stability
- Personal protective equipment
- Fall prevention and protection
- Equipment safety
- Safe handling of asbestos, PCBs and lead
- Safe use of cutting torches and arc welding
- Safe use of hand tools
- Safe blasting and explosive handling procedures
- Confined space safety practices
- Safe practice when demolishing stacks

In 1981, the Department of Labor provided a grant to the National Association of Demolition Contractors (NADC) to prepare a safety manual dedicated to safe practices for the demolition industry. Subsequent revisions were printed in 1989 and 2000. In 2004, the name of the National Association of Demolition Contractors was changed to the National Demolition Association, (NDA). The NDA's *Demolition Safety Manual* is recognized as one of the most comprehensive and user-friendly references for all demolition contractors who are dedicated to maintaining safe demolition practices for their projects.

The OSHA standards cover worker safety, and they are the only safety regulations that carry the force of federal law enforcement. Serious violations of OSHA standards can result in heavy fines and even criminal prosecution. The requirements of OSHA standards contained in 29 CFR 1926.850, Subpart T as follows are of critical importance to the demolition industry:

> 1926.850(a) requires an Engineering Survey for every demolition project. This survey must be conducted by a competent person as described in 29 CFR 1926.32(f)—"Competent person" means one who is capable of identifying existing and predictable hazards in the surroundings or working conditions which are unsanitary, hazardous, or dangerous to employees, and who has authorization to take prompt corrective measures to eliminate them.

A proper Engineering Survey also gives the demolition contractor a uniform and detailed system to inspect the project site for any problems relating to potential hazards and emergencies. The survey should also list all types of hazardous chemicals, gases, explosives, flammable materials, or similar materials that may have been used or stored on the site.

Other standards in Subpart T require a competent person to conduct a continuous inspection as the demolition work progresses to detect hazards resulting from weakened or deteriorated floors, walls, or loosened material. No employee shall be permitted to work where such hazards exist until they are corrected by shoring, bracing, or other effective means. To comply with this standard, a competent person must be on each demolition project at all times.

Although Subpart T is the standard that is specific to demolition, there are other OSHA standards that require compliance. It is safe to assume that every standard in CFR 1926 will apply to a demolition project at one time and most will apply at all times. In addition, there are many standards in CFR 1910 that apply to demolition projects, for instance, if working in an operating refinery, one would be subject to 1910.119, "Process Standards."

OSHA can also cite a contractor under OSHA Section 5(a)(1), "General Duty Clause," which requires an employer to furnish to its employees: "employment and a place of employment which are free from recognized hazards that are likely to cause death or serious physical harm to his employees..."

Compliance with OSHA Standards on a demolition project is crucial, and non compliance may become life threatening in some cases. A good safety program will not only address compliance, but will also include employee training and discipline.

C. EMPLOYEE TRAINING

OSHA addresses employee training and education in 29 CFR 1926.21, which is designed to cover all construction activities. In order for a contractor to maintain a safe environment for his or her employees, it is critical that they understand the hazards of the project and comply with safe work practices.

The Project Specific Safety Plan along with a Task Hazard or Job Hazard Analysis is essential in identifying hazards and addressing their control measures. For example, if a contractor is going to demolish some obsolete chemical process equipment, it would be the contractor's duty to ascertain the potential hazards in that work area. This can be done by having the area surveyed by competent person(s) to determine if there are any hazards present and reviewing records of what has been done to mitigate any residual hazards. In addition to the potential hazards of dangerous residual material in the equipment, it is also necessary to determine the structural stability of the equipment and any supports that will be demolished as well as the need to de-energize the equipment and disconnect utilities. Once the hazards have been identified, the demolition contractor should communicate these potential dangers to his employees and be assured that all personnel on the site understand how the work is to be accomplished in a safe manner.

Figure C.01. Safety training.

It is necessary to periodically reevaluate the hazards and re-educate the employees as the work progresses and the hazards change. Documentation of Site Specific and Task Hazard awareness training is very important. More important is the need to ensure the employees know what the hazards are and what protection is required. The last thing a contractor will want to hear if there has been an accident is "I didn't know it was dangerous."

Handling Asbestos Containing Materials (ACM) is a task that requires specialized training. The personnel safety training for working with ACM is set forth in detail by OSHA 29 CFR 1926.1101. The Environmental Protection Agency (EPA) regulations that cover asbestos are set forth in CFR Part 763, "Asbestos," and CFR Part 1, Subpart M, "National Emission Standards for Asbestos." Only employees trained and qualified shall be allowed to work with ACM.

OSHA offers ten hour and thirty hour worker training in construction safety and forty hour specific training for workers if hazardous materials are involved. The forty-hour courses are referred to as "Hazwoper Training." Special training is also required for such tasks as the operation of forklifts, aerial lifts, fall protection, and scaffold erection. Some labor unions have extensive worker training that can be a real safety benefit for both the employees and the contractor. Demolition-specific safety training is also available from independent safety training companies.

Thorough training of the demolition contractor's employees is an ongoing process. The safety conscious contractor will continue to provide required general safety training. In addition, they will provide site-specific hazard training relative to the project-specific tasks being performed and verify that such training is being done on all jobsites.

D. DRUG AND ALCOHOL POLICY

Another safety issue relating to employee behavior is a strict prohibition of drug and alcohol abuse. The contractor should have a written policy informing employees that any drug or alcohol abuse is prohibited, and the following rules should be made a part of this policy:

- There will be no drugs or alcohol allowed on a jobsite.
- An employee who reports for work under the influence of drugs or alcohol will be prohibited from work and may, in some cases, be discharged immediately.
- The employer may include a drug/alcohol testing program that may be either random checking or on a company-wide basis.
- Most drug/alcohol policies provide for termination of an employee who violates the rules.
- Some labor unions have drug and alcohol training and testing programs, but the employer is ultimately responsible for the actions of all employees.
- The employer should require mandatory drug and alcohol testing for all people involved in accidents.

Customers will frequently have their own drug and alcohol abuse policies in force, and the demolition contractor may be required by contract to enforce this policy in addition to his or her company's own policy. The project foremen and supervisors should be trained to recognize symptoms of drug and/or alcohol abuse, and they must continually be on the alert to ensure that no employee is impaired by drugs and/or alcohol abuse on the jobsite.

E. PERSONAL PROTECTIVE EQUIPMENT

Personal Protective Equipment (PPE) is an extremely important component of protecting the demolition worker. PPE is really a two-part safety requirement:

Part 1: There are certain items of PPE that are required for all projects as a bare minimum:

- hard hats
- safety glasses

- substantial boots, with safety toes
- gloves
- hearing protection
- high-visibility vests

Part 2: Depending on the demolition tasks being performed, the following items may also be required:

- full face shields
- safety toed boots with metatarsal protection
- respiratory protection
- fireproof clothing
- life preservers for any project around or over water
- fall protection devices such as body harnesses
- chemical resistant gloves and outerwear
- personal industrial hygiene monitors

There can be some difficulty for the contractor to enforce the everyday compliance with the PPE requirements for a given project. Workers and their supervisors may not always follow the PPE requirements established for the job. Therefore, the demolition contractor and/or the job superintendent must take severe action against any employee and the employee's supervisor for failing to use the PPE required for a particular phase of the demolition project. Failure to use the proper PPE is a violation of OSHA standards and can result in serious safety consequences as well as incurring monetary and criminal penalties.

F. JOBSITE SECURITY

Protective measures to maintain a safe and secure demolition jobsite include fencing, signage, pedestrian walkways, scaffolding, enclosures, temporary lighting, and moveable concrete barriers. When working in urban or congested areas, the demolition contractor may also use safety and security personnel to help protect the public from potential dangers created by falling demolition debris, trucking operations, and the movement of heavy equipment.

Also an integral part of jobsite safety is the protection of the workplace from illegal activities. As an example, there are a number of issues that should be addressed when protecting a worksite from dangerous activities:

- *Illegal entry*—Unauthorized people can be injured when "exploring" a demolition site, resulting in serious consequences for the demolition contractor.
- *Vandalism*—This is such a broad category of security failure that it is impossible to include all of the possibilities. Serious damage to equipment is probably the most significant problem.
- *Arson*—All too commonly "strange" fires appear at demolition sites. For example, piles of combustible debris can be a very tempting target for the serious vandal.
- *Theft*—Although statistics are difficult to quantify, theft is a serious problem for the entire construction industry, and demolition contractors are no exception. There are instances of large pieces of heavy equipment, such as excavators, being stolen from jobsites and transported hundreds of miles away for illegal sales. Theft of non-ferrous metals, especially copper and aluminum, from electrical equipment like transformers can pose both electrical hazards and result in spills of PCBs.

Figure F.01. Safety closure with appropriate signage.

Figure F.02. Erecting safety barricade.

Fencing with locked gates will keep out the casual perpetrator; however, in some areas, effective security will require a guard service and perhaps one with dogs to protect the demolition site during nighttime hours.

Site security is an expense to the cost of doing work and must be taken into account when estimating a demolition project.

G. STUDY QUESTIONS

1. With a good safety record, what kind of benefits can a company expect?

2. What type of documentation is required by OSHA on demolition projects and needs to be completed by a competent person?

3. What types of rules must be included in a contractor's drug and alcohol policy?

4. What are the two parts to the OSHA requirement regarding Personal Protective Equipment?

5. In jobsite security, what are four dangerous activities that often occur on demolition project sites?

CHAPTER 9

DEMOLITION EQUIPMENT

A. INTRODUCTION

Prior to the early 1960s, most equipment used for demolition was ordinary construction equipment such as loaders, trucks, and cranes. As the demolition industry responded to the demands of the interstate highway construction and the urban renewal programs, the equipment manufacturers began to produce attachments designed for demolition. One of the best examples of this was the development of the demolition bucket designed for front-end loaders. This bucket had a top mounted clamping device to enable the operator to grab and hold demolition debris as well as to give the machine an extended reach capability for wrecking low-rise buildings. Today, a demolition contractor can purchase many different types of equipment and attachments that are specifically designed for demolition work. Many demolition contractors have designed and built their own specialty tools for particular applications. This chapter describes primary heavy equipment, support equipment, and specialty equipment items for demolition.

B. PRIMARY HEAVY EQUIPMENT

B.1. Excavators

The most common method of demolition now employed by the industry in North America is the use of crawler excavators equipped with a grapple or bucket and thumb attachment. These machines come in sizes of five tons to over one hundred tons gross weight and can use several different types of booms and attachments to demolish a wide variety of structures. Many excavators are equipped with a "quick mounting and release" mechanism that permits the operator to change from one attachment to another in a matter of minutes.

Figure B.01. Excavator with concrete pulverizer.

Figure B.02. Excavator with bucket and thumb.

With the addition of another boom section known as a "third member" in the excavator boom, heights of over one hundred feet can be reached. These excavators are in a class of machine known as "Hi-Reach Demolition Machines" and are in the thirty ton to 150 ton class. Using a shear, concrete pulverizer, concrete cracker, or other attachments, a ten-story building can literally be "chewed" apart. These machines are replacing, to a large extent, the use of cranes to reach buildings and other structures of up to ten stories.

Figure B.03. High-reach demolition machine.

Figure B.04. Excavator with shear.

Figure B.05. Excavator with grapple.

Figure B.06. Mini-excavator with concrete cracker.

A very important point in working with a high-reach demolition machine is to ensure that the machine is always operating over a very stable and reasonably level surface. The operator must have a good field of vision and usually works with a spotter who can signal as necessary, or a video camera can be mounted on the boom tip to give the operator a clear view of the work. It is also important to operate the machine in such a manner as to keep the angle of the main boom within the limits of the machine capacity to prevent unbalancing. Water spray nozzles can be installed near the attachment to control dust and to allow the operator to see what he or she is doing in real time.

Smaller excavators are used extensively for floor-by-floor demolition techniques and for a variety of selective demolition requirements.

Attachments commonly used with excavators include the following:

- Shears for cutting metal and concrete
- "Crackers" used for fracturing heavy concrete
- Impact hammers to break concrete not easily fractured by jawed attachments
- Universal processors for breaking concrete into small pieces and shearing rebar
- Breakers for breaking heavy concrete
- Grapples for demolition and handling bulky debris
- Magnets

Figure B.07. Excavator with impact hammer.

B.2. Cranes

Without the use of modern cranes, many types of demolition projects would be far more difficult. There are no cranes manufactured strictly for the demolition industry. Cranes using wrecking balls must be classified as "duty cycle" cranes and allow the ball to free fall. Such tasks may include the following:

- Demolishing a building using a wrecking ball that may weigh 1,500 to 10,000 pounds
- Lifting small equipment for use in high-rise demolition and interior demolition
- Demolishing a building with a clam shell bucket (to a large extent, the clam bucket has been replaced by the excavator and grapple attachment)
- Lifting down sections of structures and equipment

Cranes of all types are rated by their lifting capacity with a defined length of boom and a radius of lift. For example, a crane rating of one hundred tons may be defined as the load limit of that particular crane with sixty feet of boom at a distance of fifteen feet from the center of gravity of the crane to the point of attachment of the load. Every crane must have a lifting chart that shows safe operating ranges available to the operator so that overloading can be prevented.

Lift capacity for a crane also varies depending on the position of the boom in relation to the tracks or outriggers that support the machine. For a tracked crane, the highest capacity is over the front of the tracks. As the boom is moved to the side, the crane is not able to lift as great a load without tipping. Other factors that should be taken into consideration when planning a heavy lift are the effect of wind, any amount of swing in the load, the hoisting speed, and the impact of stopping a lift abruptly. A 50% safety factor is advised for all demolition lifts because of the influence of these factors coupled with the unknowns that complicate the weight estimated for demolition materials to be lifted. It is always better to have excess lift capacity because once a demolition lift is started it is usually not possible to reconnect the load to the structure.

The variety of cranes commonly used in the demolition of buildings and other structures may include the following generic types:

Mobile Cranes—Mobile cranes are mounted on a chassis supported by rubber tires and can have three to eight axles. These cranes are widely used in the demolition industry and come in a variety of configurations and capacities. The mobile crane used for demolition has a lattice-type boom. Mobile cranes with hydraulic booms are used only for lifting tasks. The ease of moving these cranes from site-to-site makes them ideal for small demolition projects.

Crawler Cranes—These cranes are mounted on a platform known as a "car body" and supported by two crawler-type track assemblies. Crawler cranes are usually used for long-term projects where the cost of mobilization is relatively minor compared to the overall project cost. These cranes are fitted with lattice-type booms and most are in the 100 to 250 ton capacity range. The larger crawler mounted cranes, because of their size and weight, must be partially dismantled before they can be moved over the highways.

Rough Terrain Cranes—These cranes are mounted on frames supported by four or more large tires that are designed to maintain the stability of the crane when traveling over uneven ground. This type of crane is ideally suited to provide services such as lifting equipment at industrial sites and removing structures and equipment in rough terrain.

Tower Cranes—The familiar tower-type cranes found on most high-rise construction projects are also used for handling demolition materials and lifting small equipment when demolishing high-rise buildings at locations where explosives cannot be used.

B.3. Loaders

Both track loaders and rubber tired wheel loaders are used in the demolition business with each machine having unique capabilities. Loaders are used primarily for moving demolition materials and loading trucks or drop boxes. However, loaders are also used to wreck low-rise buildings. Small loaders, mostly of the skid-steer type, are extensively used for moving materials, cleanup duties, and for demolition with small shears and hammers.

B.4. Concrete Crushing

Crushing concrete and asphalt for recycling is now a common practice throughout the demolition industry and there are two primary types: jaw crushers and impact crushers.

Jaw crushers are the most simple and are generally used when crushing concrete to a size of not less than three inch minus (meaning that all products will pass through a three inch screen). The mechanics of a jaw crusher operation are simple. An electric or diesel motor provides power to a moving plate fixed to an eccentric shaft that is thrust back and forth against a fixed plate. Both plates have replaceable wear surfaces. A large flywheel stores energy as it rotates, and the material to be crushed is gravity fed into the opening where it is crushed to a size that will permit passage to an off-loading conveyor at the bottom of the opening.

Figure B.08. Crawler crane with wrecking ball.

Figure B.09. Rough terrain crane.

Figure B.10. Hydraulic crane.

Figure B.11. Tower crane for floor-by-floor demolition.

Figure B.12. Wheel loader with grapple type bucket.

Figure B.13. Skid steer with grapple bucket.

Figure B.14. Track loader with grapple bucket.

Impact crushers are either electric or diesel powered and break concrete by impacting it with blow bars that are mounted to a spinning shaft inside a chamber. The blow bars pound the concrete pieces against the chamber walls. The discharge opening can be set to crush concrete down to one and one half inch minus by continuously recycling the oversize material back into the impact chamber. The blow bars resemble "swinging hammers," and the chamber is lined with replaceable wear plates similar to that on the jaw crushers.

The common enemy of both types of crushers is the presence of reinforcing steel (rebar). The normal procedure is to cut the rebar before it is fed into the crusher so that not more than six inches of rebar protrudes from the concrete being crushed. A transverse magnetic conveyor picks up the pieces of rebar from the discharge conveyor and drops it into a pile. Both types of crushers may be mounted on either a trailer chassis or on a crawler chassis.

There are occasions where a small mesh product, such as three-fourths inch minus is required. Small products may be achieved by passing the previously crushed product through a cone crusher. The cone crusher is simply a heavy steel cone rotating on an eccentric shaft within a cylindrical tub and is set so that it only allows material crushed to the desired setting to pass through.

Depending upon the project, one or several conveyors may be used to stack the crushed material or to feed it into screens for size separation. For further size reduction, material can be fed into a cone crusher.

Figure B.15. Typical crushing operation.

B.5. Robotic Equipment

For certain specialized demolition tasks, the robotic machine is the ideal piece of equipment. These machines resemble mini-excavators without a cab for an operator. Instead, the operator stands safely away from the machine and performs any number of tasks. These robotic machines can be fitted with shears, hydraulic hammers, grapples, concrete pulverizers, hydraulic drills, and excavator-type buckets with thumbs. They are ideally suited for operating in a wide variety of environments such as:

- Cement kilns—for breaking out cement accumulations and refractory brick
- Hi-rise demolition—for breaking concrete and shearing rebar and light steel members
- Smoke stacks—for operating off of a crane supported platform to break apart concrete and/or brick
- Demolition of portions of structures where floor load ratings are unsafe for larger equipment
- For demolition of structurally unsafe buildings—operator can be at a safe location in the event of an unplanned collapse
- For operation in hostile environments such as radioactive areas, high temperature conditions, and areas of known chemical and explosive hazards

Figure B.16. Robotic machine with concrete cracker. Operator using joystick to control machine functions.

Figure B.17. Robotic machine using hammer to wreck stack. Operator is on the ground, crane is used to position the machine.

Figure B.18. Large wheel loader with fork attachment moving slabs of concrete.

B.6. Forklifts

There are several varieties of forklifts used for material handling at demolition sites. Following are forklifts that are typically found on demolition projects:

- Rough terrain, four-wheel drive forklifts with an extendable boom for handling material from the ground to elevated work areas. For example, to place tools on the second or third floor of a building.
- Pneumatic or hard rubber tired standard forklifts for moving and loading materials such as pallets of used bricks or bundles of salvaged timbers.
- Skid-steer loaders with removable forks for operating in buildings or on-site.

C. SUPPORT EQUIPMENT

There are numerous machines used in the demolition process for everything from access equipment to specialty cutting applications. The more common examples are below:

C.1. Access Equipment

- *Man Lifts*—Four-wheel drive with telescoping booms allowing one or two workers to commonly reach heights from forty to over one hundred feet

Figure C.01. Man lift.

Figure C.02. Man basket.

- *Scissor Lifts*—Four-wheeled platforms for accessing heights typically from ten to forty feet from a flat, level surface providing a stable work platform for removing building materials such as siding materials and piping
- *Scaffolding*—The oldest and most common systems for providing long-term work platforms either inside or outside of structures
- *Crane Supported Man Baskets*—Although not usually used if other methods are available, a two to three person man basket can be lifted with a crane to gain access that would not otherwise be possible or safe
- *Moveable Scaffold System*—Capable of lifting equipment such as skid-steer loaders and compressors to working floors

Figure C.03. Movable scaffold platforms.

C.2. Concrete Breaking and Cutting Equipment

- *Compressors*—There are many types and sizes of air compressors used to supply air for breaking and cutting concrete. They are either gasoline/diesel powered or electrically-powered, and sizes typically range from eighty cubic feet per minute (cfm) trailer mounted units to over nine hundred cfm mounted on trailers or skids.
- *Pavement Breakers*—Often called "jackhammers," these usually operate on eighty to ninety psi air pressure from a portable compressor and use an impact action to break concrete or rock.
- *Concrete Slab and Wall Saws with diamond impregnated blades*—Most common are the walk-behind slab saws to cut floors and slabs. Wall saws are moved along tracks that are bolted to concrete or masonry walls and can cut sections out of walls such as to cut an opening for a door.

Figure C.04. Pneumatic pavement breaker.

Figure C.05. Walk-behind concrete saw.

- *Concrete Coring Drills*—Are used for cutting holes in concrete walls and slabs; mostly used in selective demolition work. May be powered by fuel, electricity, or air.
- *Concrete Splitters*—These hydraulically operated, handheld tools use the wedge principle to split large sections of concrete. Simply put, a sliding wedge moves between two fixed wedges inside pre-drilled holes, and a hydraulic pump is used to exert enormous force on the moveable wedge, thus spitting the concrete much as a wedge is used to split wood.
- *Concrete Splitting Compounds*—These are mixtures of specialized cement and clays that are mixed with water and placed in pre-drilled, holes. A chemical reaction causes the material to expand as it becomes hard, and this exerts enormous pressure, causing the concrete to crack (same principle as the hydraulic splitters described above).
- *Rock/Concrete Drill*—Looks much like a pavement breaker but it has both a rotary and hammering action and is used to drill holes in rock or concrete.
- *Rivet Buster and Chipping Hammers*—These are handheld tools, similar to a small jackhammer, used for a variety of small concrete breaking and chipping jobs.
- *Winches*—Powered by fuel, electricity, or air, winches are used for lifting and pulling equipment and materials inside of buildings.

C.3. Metal Cutting Tools

- *Oxy-Acetylene and Oxy-Propane Cutting*—Are perhaps the most common method of cutting steel in the demolition industry. Acetylene is somewhat faster, but not as safe and is more expensive than propane.

Figure C.06. Concrete splitter.

- *Abrasive-Type of Circular Saws*—Portable, handheld gasoline, pneumatic, or electrically-powered saws for cutting pipe and small structural steel or sheet metal.
- *Plasma-Arc Torches*—Expensive but useful devices for cutting stainless steels and other exotic metals, which require higher temperatures or in areas where sparks must be minimized. An inert gas or compressed air is fed at high pressure through a nozzle where it passes through an electric arc that turns some of the gas into plasma and melts the metal being cut.
- *Burning Bars*—Are long tubes (about 10 feet), which are handheld and contain various assortments of iron, magnesium, and aluminum rods. Air is forced through the tube and an electric spark or an oxy-fuel torch is used to ignite a flame. An exothermic reaction results in temperatures up to 8,000 degrees Fahrenheit at the tip. They can burn through most anything including metals and concrete.

Figure C.07. Chipping hammer.

Figure C.08. Steel cutting with long torch.

- *Reciprocating Saws*—Handheld, electrically-powered hacksaws used for cutting pipe, cable, and anything that is located in an area where torches are not allowed.
- *Small Shears*—Handheld, hydraulic, or electrically-powered shears are used for cutting small metal sections and reinforcing steel up to one inch. Other styles are used for cutting steel and copper cable.

D. DEMOLITION TRUCKING

Trucking, of various types, is required for most demolition activities from transporting equipment to and from the project to hauling all materials that are produced in the demolition process. In most locales, a typical demolition project will include the following examples of trucking:

- Transporting heavy equipment, such as excavators and loaders, on lowboy trailers to and from the site
- Transporting support equipment, such as forklifts and skid-steer loaders
- Bringing fuel to the site
- Providing on-site water trucks for fire protection and dust control
- Hauling hazardous materials to specialized disposal sites

Figure D.01. Lowboy configured to meet department of transportation requirements.

- Hauling demolition debris to a landfill
- Hauling broken concrete, asphalt, and masonry to a recycle center
- Hauling salvaged scrap metals to a recycle center
- Hauling other salvage, such as useable equipment or salvaged timbers, to a buyer

Sizes of hauling units and the number of axles needed are governed by the weight restriction of the state and local governing agencies. It is standard practice to cover the loads with tarps to prevent dust and small pieces of material from blowing off of the trailer.

The most common method for hauling demolition debris, such as wood, drywall, roofing, and plastic materials, to a landfill is to use high side end dump trailers pulled by a minimum of three axle tractors. Such trailers can haul from fifty to one hundred cubic yards, depending upon the density and weight of the material being hauled. Roll off boxes from twenty to forty cubic yards are also extensively used for this type of hauling. Walking floor trailers, which off-load the material with a conveyor-like action on the bottom, are also used for hauling debris that has been crushed to small pieces.

Concrete, asphalt, and masonry is usually hauled to a recycle plant in twenty to twenty-five cubic yard (cy) end or side-dump trailers pulled by a three axle tractor. Three axle, ten to twelve cy dump trucks are sometimes used for short haul requirements. Other combinations of tractor and trailers are also used for hauling these materials.

Scrap metals may be hauled in any of the above type of trailers and drop boxes except for the live bottom type.

Figure D.02. Loading high volume debris trailer.

Figure D.03. Loading side-dump trailer.

Figure D.04. Roll-off drop box.

Figure D.05. Transport for crawler crusher.

Figure D.06. Vacuum truck.

Figure D.07. Water truck.

Earth excavation and backfill materials are hauled in a wide variety of tractor-trailer combinations that are common on all types of construction projects.

On large industrial projects, off-highway trucks that can haul loads of fifty tons or more can be used if the mobilization cost is not prohibitive.

On large scrap projects with railroad access, rail cars can haul prepared scrap directly to the mills.

E. SPECIALIZED EQUIPMENT

Demolition contractors, over the years, have shown a remarkable ability to customize tools for special demolition purposes. Early grapples evolved from efforts to equip the excavator with a tool similar

Figure E.01. Pressure washer cleaning contaminated concrete. Higher pressure washing equipment of up to 40,000 psi is available.

Figure E.02. Loader with custom boom.

Figure E.03. Thumper for breaking slabs.

Figure E.04. Custom bridge box to catch broken concrete.

Figure E.05. Floor scraper.

to a clam bucket commonly used with cranes. Various types of mining equipment were customized to serve the needs for drilling and breaking concrete foundations. Custom-built chutes are used to convey demolition materials to a truck or drop box.

Some projects have a need for custom-built devices that could be used for protecting utilities and equipment or structures. Protection for fire hydrants, transformers, and covers for floor openings are typical examples.

F. STUDY QUESTIONS

1. What type of equipment is replacing the use of cranes to reach buildings and other structures of up to ten stories?

2. What equipment is best utilized to lift small equipment for use in high-rise demolition and interior demolition?

3. What kind of crane is supported by four or more large tires, which are designed to maintain the stability of the crane when traveling over uneven ground?

4. What are the unique capabilities distinguishing the track loader and the rubber tired loader?

5. What are some common examples of access equipment?

6. What governs the size of the units, and the number of axles per unit, needed to haul heavy equipment?

CHAPTER 10

MATERIAL HANDLING AND RECYCLING

A. INTRODUCTION

Efficient material handling and recycling is as important a process as the actual demolition of a structure. In fact, the major cost of a demolition project is for sorting, recycling, hauling, and disposing of the products of demolition. Demolition contractors who plan their work so that they will recover the maximum amount of marketable materials will, by so doing, increase their income by the net value of the salvage and reduce the amount of material that must be hauled to a landfill.

B. MATERIAL HANDLING AND SORTING

Before the 1990s, most demolition debris such as wood, drywall, insulation, masonry products, concrete, asphalt, and less valuable metals were taken to landfills. Construction and demolition debris is usually referred to as C & D debris and consists of any kind of waste material that is generated in either the construction or demolition of buildings. As the public became more aware of the problems related to disposing of C & D debris in landfills that had diminishing capacity, C & D debris separation for recycling became more important. Hauling of C & D debris to landfills is often the largest single cost of a demolition project.

As the environmental problems such as leaching of heavy metals and chemicals into groundwater from landfills became more apparent, governmental agencies created regulations that required significant improvements to the operation of landfills and the closing of substandard landfills. As a result, the cost of landfill disposal has greatly increased over the last thirty years. The increased cost of disposal, and in some cases the requirement for mandatory recycling, affects the way a demolition contractor demolishes buildings and sorts the materials. For a well-managed demolition project, the sorting of materials produced in the demolition process has become a very important sub-task of the overall demolition process.

An effective way to sort demolition materials is to demolish the building or structure in such a manner that the various classes of materials can be efficiently separated. For example, assume an apartment building having a structural concrete frame and floors and interior construction of light wood frame construction is to be demolished. The demolition contractor can "gut out" (demolish) the interior partitions, ceilings, and any non-concrete materials, pushing these materials out of the building to be recycled separately or disposed of at a landfill. The remaining concrete structure can then be demolished and the concrete processed and crushed to produce a useable product either for backfill or for sale to an off-site customer.

Another common practice in debris sorting is accomplished during the demolition process by use of an excavator equipped with a grapple. In this process, the excavator operator can separate the construction materials as the demolition progresses. In the process the operator will place landfill debris in one pile, masonry in another, and scrap metals in another. This process is most effective when the building is primarily constructed of wood and masonry materials. For steel and concrete construction, this process works well when two excavators are working together. One is usually equipped with either a shear or concrete cracker to cut or break the building components while the second excavator is equipped with a grapple to grab and place similar materials in separate piles. Some contractors may also use electromagnets to pull ferrous metal from the debris.

C. DEBRIS REMOVAL AND DISPOSAL

In the demolition industry, material handling is a broad term that can refer to many types of activities from delivering scaffolding to a jobsite to trucking debris to a landfill. Although material handling includes the handling of recycled materials, this chapter treats recycling as a distinct and separate operation.

Handling Landfill Debris—Landfill debris generally refers to those materials that have little or no net value as recycled materials. These materials include small dimension lumber, drywall, insulation, carpeting, plastic materials, roofing materials, and other materials that are not typically recycled due to lack of ready markets or high processing costs. These items are also referred to as "soft demolition" materials as opposed to materials such as concrete, masonry, and metals. During the demolition process the soft demolition materials are ground or broken up to reduce their bulk volume so that more materials can be loaded into a truck or roll-off drop box. On smaller projects, the grinding is commonly accomplished by using a track loader or dozer to run over the material until it is reduced in size to permit efficient loading and hauling.

Figure C.01. Reducing wood bulk with dozer.

On larger projects, it is often feasible to bring in a tub grinder to shred the material into small pieces to achieve maximum loads. Once the material is sufficiently broken up, it is loaded into either large volume dump trucks or into roll-off drop boxes with either an excavator equipped with a grapple or a loader equipped with a grapple type of bucket. Dump trucks typically range up to one hundred cubic yards of haul capacity. Roll-off drop boxes vary from twenty to forty cubic yards. In many areas, the loads are required to be wet and covered with a tarp before entering the highway to keep all dust and debris within the truck or container.

At landfill sites that charge by weight, the trucks are weighed and then proceed to the designated area to dump their load. Empty trucks are re-weighed on their way out of the landfill to determine the cost of disposal based on the weight delivered. For example if a truck delivers fourteen tons of debris and the disposal fee is $50 per ton, then the load will cost $700. In some cases, the disposal fee is calculated on the basis of volume. In Europe, Canada, and some areas of the U.S., the disposal fee is referred to as the "tipping fee."

If roll-off drop boxes are used, they are usually delivered to the site and picked up by the firm that owns the boxes. Disposal fees are either included as a part of the fee for the use of the box or calculated separately when the net weight is determined at the landfill scale.

For interior demolition, the materials are moved out of the building using a combination of hand labor, wheeled hand-carts, skid-steer loaders, chutes, lifts, service elevators, or empty elevator shafts for multi-story buildings. In some situations, small conveyors are used to move materials out of the work area.

The following photos are examples of handling demolition materials at typical projects.

Figure C.02. Loading debris.

Figure C.03. Loading steel scrap.

Figure C.04. Removing and handling roofing debris.

D. RECYCLING

As an industry, demolition contractors have been at the vanguard of recycling. Sale of recycled materials is an income generator, and if a building material product or equipment item can be removed and sold for more money than it cost to remove it, then the demolition contractor would take that action. At times the demolition contractor must pay to dispose of material for recycling. If the cost to haul and dispose of the material at a recycling facility is lower than the cost of disposal at a landfill, the contractor has still reduced his or her overall project cost.

As markets and regulations governing disposal of C & D materials have changed, the emphasis on recycling has changed. The following examples represent some of the more important aspects of recycling building components:

Scrap Steel and Iron—For centuries, scrap steel and iron have been recycled. Their value is a significant component of the income stream on many demolition projects. Scrap steel includes such items as structural steel, pipe, sheet metal, fabricated products, and reinforcing bars. Scrap cast iron is mostly in the form of large machinery components. Scrap stainless steels are especially valuable because of their chrome and nickel content, bringing much higher prices than common steel scrap. Scrap steel from demolition projects is separated into two major categories—#1 Heavy Melt and #2 Scrap, consisting of sheet metals and other light weight items. Also, steel and iron scrap is shipped either as unprepared, meaning any size that will fit in the truck or railcar, and prepared scrap, which is cut into pieces usually less than 5 feet long and 18 inches wide. The unprepared scrap requires that the scrap yard perform additional sizing before shipping to a mill, while prepared scrap can be shipped directly to a mill, usually for a higher price to the demolition contractor.

Figure D.01. Sorting and sizing steel scrap.

Steel and iron construction materials are usually separated during the sorting process with an excavator equipped with a grapple attachment or a bucket and thumb or an electromagnet. They are either sent to the scrap yard in the contractor's trucks or loaded into containers furnished by the buyer. Depending on the particular project, the steel and iron may be shipped in large pieces for further processing by the buyer or cut into smaller pieces with either shears or torches to a size acceptable to the mills' specifications for melting. The price obtained for the scrap steel may be increased by cutting the steel to the mill's specification and separating the steel by product and chemistry.

Figure D.02. Processing steel scrap in burn yard.

Non-Ferrous Metals and Alloys—Metals such as copper, aluminum, lead, brass, titanium, and numerous alloys are nearly 100% salvaged on demolition projects. Copper wire is the most common non-ferrous material that is salvaged and sold to recyclers. The wire is either pulled from conduits before demolition begins or is removed from the waste stream as demolition progresses. Contractors may strip the insulation from copper wires to obtain the highest price or send it to the scrap yard for further processing. These metals are used in many types of process equipment, piping, and electrical systems. Most demolition contractors will try to remove and separate these metals early in a project since they represent high value recyclables. Items such as piping and vessels containing these metals are usually sorted out of the waste stream as the demolition progresses. Both skid-steer loaders and larger front-end loaders are usually the preferred machines for moving these metals to stockpile or loading them into roll-off drop boxes and trailers. Excavators with grapples or tank shears and larger loaders with grapple buckets are used to handle bigger items such as tanks and vessels.

Concrete and Concrete Masonry Unit (CMU) Products—Serious recycling of concrete and CMU began in the early 1990s and represents the most significant tonnage of demolition recycling. During the demolition, sorting, and preparation process, rebar is separated from the concrete pieces to be handled and sold as scrap steel. In recent years, it is has been estimated that annually over 140,000,000 tons of concrete was crushed and recycled.

Most concrete is recycled in one form or another. It is either crushed by concrete crushers into a uniform size or it is broken into chunks usually smaller that twelve inches. Both the chunks smaller than twelve inches and the crushed concrete can be used as backfill material for basements, excava-

Figure D.03. Material handler with magnet loading steel scrap.

Figure D.04. Processing and separating rebar from concrete prior to loading.

tions, etc. Concrete and CMU crushed to a minus six inch size or smaller is also often used as roadbed material, base courses for parking lots, and in some cases, as aggregate for new concrete.

Crushing can take place on-site using portable crushers, or taken to nearby permanent crushing facilities.

Depending on the size of the crusher, concrete is broken into pieces small enough to be fed into the jaws of the crusher. This is usually done by a concrete pulverizing attachment or hydraulic hammer. Rebar protruding from the concrete may have to be trimmed first with torches or shears. Rebar remaining in the concrete is expelled by an electromagnet as the crushed product is conveyed to a stockpile and is recycled separately as scrap steel. Non-ferrous metal, wood, and other debris that goes through the crusher is typically separated by manual labor as the crushed material travels down a conveyor belt.

CMU, often known as either concrete blocks or cinder blocks, does not have nearly as much rebar and is, therefore, much less of a problem to handle. These materials are commonly mixed with the concrete for crushing and recycling. Masonry, such as unsalvageable brick and tile, is not as desirable as concrete and CMU for recycling, but can be used as backfill if specifications permit.

Brick—In some geographic areas, used brick, especially those fired at higher temperatures, is considered architecturally desirable material and is salvaged for resale. Softer brick or common brick is generally too soft or porous to face new buildings and has little resale value.

Asphalt Paving—Asphalt paving is recycled in many areas for reuse in new asphalt for roads and parking lots. Many states accept some percentage of recycled asphalt as aggregate in new highway asphalt mixes. Private projects frequently use recycled aggregate asphalt due to the savings in paving cost that results from the recycled content.

Wood Materials—Large dimension timbers have always been a popular item for salvage and resale. These timbers can be milled into many new products. One popular re-milled product is hardwood flooring where the stability of the old growth wood in these timbers is highly prized. Small dimension lumber, less than four by six feet, is seldom valuable enough to salvage for resale because of the high labor cost involved. Rare types of wood, such as old growth heart pine, oak, redwood, and cherry, are always in demand and are salvaged whenever possible. In some cases, wood debris without lead paint contamination can be ground up and used for mulch, animal bedding, or burned as fuel where appropriate facilities are located within a reasonable hauling distance. Trees and brush that are removed as a part of a demolition project are also good candidates for chipping and use as mulch.

Machinery and Equipment Recycling—Some demolition projects, mostly those of an industrial nature, will contain useable equipment such as electrical switch gear, instrumentation, air compressors, pumps, valves, boilers, vessels of all types, and a wide variety of specialized process equipment or miscellaneous items that have a potential for reuse. For valuation of specialized equipment, the demolition contractor will often consult with potential buyers as to the value of the equipment and then determine the cost of removal and shipping. Net value is the amount left after costs for removal and handling have been subtracted from the sale price. For certain items, the expertise of a specialist to aid in the salvaging effort is well worth the expense. An example would be a turbine generator that requires special procedures to avoid damage during removal and handling.

Architectural Features—Items such as stained glass windows, stone, terra-cotta building components, various examples of period architecture, and other unique construction features are popular salvage and resale items. It is sometimes necessary to employ craftsmen with special skills to carefully remove these items for salvage and resale. Occasionally, demolition specifications will require the contractor to carefully remove such items and turn them over to the owner.

Excavated Earth—If a demolition job will require the excavation and removal of earth materials from the site, the demolition contractor may find a buyer or haul it to a landfill for use as cover material.

The following photographs depict some of the more common recycling operations:

Figure D.05. Salvaged timbers stacked for shipment.

Figure D.06. Salvaged brick.

Figure D.07. Salvaged boiler tubes.

Figure D.08. Concrete crushing operation.

Figure D.09. Loading steel scrap for shipment to mill.

E. RECYCLING HAZARDOUS MATERIALS

In addition to the above recyclable materials, there are a few hazardous materials that can also be recycled. For example, even though there is a miniscule amount of mercury in an older, single fluorescent lighting tube, when hundreds of tubes are removed and processed, the mercury can be recovered and processed for reuse. Also, refrigerants from some cooling systems can be removed and processed for reuse.

Many types of petroleum products, often referred to as POLs, are pumped from their containers and reprocessed for use as fuel.

Some types of instrumentation and lighting devices such as exit signs contain tiny amounts of radioactive materials. The handling and chain of custody of these devices are regulated and in most cases they can be returned to the manufacturer for disposition. The general regulations set forth in 10 CFR Parts 1-171 detail how these devices are to be handled.

F. SUSTAINABILITY AND THE FUTURE OF RECYCLING

The world of recycling is rapidly changing. Drywall, carpeting, and roofing are not commonly recycled in 2010, even though there is pressure to do so. Although the technology to recycle these products exists, the structure of businesses to facilitate a full system of market exchange from project to end user has not yet matured in most of the country. Some areas in Canada already require drywall recycling. Future regulations will likely force the demolition contractor to recycle some materials even though it would be cheaper to simply dispose of them in landfills.

A trend in construction that is helping to move recycling forward is an interest in "green" building practices. While "green construction" is not always well defined, several national and international organizations have been promoting the design, construction, and operation of buildings that deliver high performance with lower levels of environmental impact. Through sustainability or sustainable design and development that minimizes waste and imposes the least environmental cost, it is hoped that "green buildings" will allow society to meet the needs of the present generation without compromising future generations' needs.

Several of these organizations promote sustainability through the certification of buildings. The certification programs provide a multi-step process. They typically accredit professionals who guide the design and construction process and publish guidelines to promote sustainability and high performance in building design, construction, and operation. In addition, they provide third-party certification that the entire process meets established guidelines or standards. Two of the most prominent programs that promote sustainability through certification are Leadership in Energy and Environmental Design (LEED) and the Green Globes system.

The LEED rating system was developed by the U.S. Green Building Council (USGBC) in the 1990s. Its initial focus was commercial and institutional projects, but during the 2000s the program went through a number of revisions. In addition to new construction and major renovation of commercial or institutional buildings, LEED currently offers certification of homes, neighborhood developments, commercial interiors, core and shell construction, and operation of existing buildings.

Green Globes was established in Canada during the 1990s. In the United States the Green Globes system is promoted by the Green Building Initiative. Although it initially provided ratings for existing buildings, during the 2000s Green Globes was expanded to include a wide range of projects that encompass new construction, renovation, and building operation. Green Globes delivers much of the assessment and rating process through online tools and is currently working with the American National Standards Institute (ANSI) to establish their guidelines as a consensus-based standard.

The certification programs that promote sustainability have complex protocols, procedures, and documentation requirements that must be followed by designers and constructors. Most of these procedures do not apply to the demolition process. Those that do apply to demolition are intended to promote reuse and recycling of materials. When implemented, these sustainable practices can help minimize the cost of debris disposal and will encourage owners to hire demolition contractors familiar with the documentation required by the certification protocols. By becoming familiar with green certification procedures commonly used in their work territory, demolition contractors can obtain a marketing advantage as well as the potential to be paid for diverting debris from landfill disposal.

Demolition Related Elements of "Green" Certification—Demolition contractors can help obtain points toward certification levels by working to preserve significant portions of an existing building for reuse. Maintaining facades, structural components, and interior non-structural components during the demolition process can all contribute certification points depending on the percentage of the building that is preserved. The percentage preserved is usually measured by area or volume. These point categories promote the use of selective demolition but may require the demolition contractor to be especially diligent in controlling dust through the use of negative air pressure containment or filtered air circulation that meets the indoor air quality requirements found in other parts of the certification process.

Certification also promotes the reduction, reuse, and recycling of demolition debris. Depending on the percentage of debris that is diverted from the landfill, points are awarded for materials that are separated and reused or recycled. Percentages are calculated either by weight or volume. Some programs allow materials to be reused for site development and landscaping. In addition, some programs give credit for debris fines that can be used as alternative daily cover at landfills or wood materials that can be used as fuel. Land clearing debris is generally excluded from these credits. It should be noted that a minimum of 50% reduction in debris disposal is generally required to begin earning certification points.

Prior to awarding certification points, the programs require a report that describes the percentage of debris generated by the project and the percentage of debris diverted from the landfill. The demolition contractor should be prepared to supply a debris management plan prior to beginning the project.

Demolition debris disposal diversion can be accomplished through material separation at the project site when space, time, and resources allow. When utilizing on-site separation, the demolition contractor should be prepared to provide an estimate of the quantity of debris to be generated by the project and the level of diversion they will provide. Upon completion of the project, a report showing the quantity of debris by category that has been diverted from the landfill is submitted to document that the required debris disposal diversion has been successfully achieved.

On-site separation is not practical on some projects. In those cases it is possible to send mixed debris to facilities that provide off-site separation and recycling. Upon request, off-site separation facilities will generally provide an appropriate report outlining the percentage of debris coming from a project that they successfully diverted.

Other green certification categories allow points to be earned through redevelopment of urban sites and brownfield sites. Demolition contractors may benefit from these activities because such projects promote property recycling and frequently require the removal of obsolete buildings or facilities. Reuse and recycling of materials not generated on the project site is also promoted by certification point categories. As green building certification becomes more popular, these opportunities to achieve certification through building product reuse and use of items with recycled content are likely to support higher market prices for materials salvaged from future demolition projects.

G. MARKETING SALVAGE

Marketing recycled products from demolition projects is generally a straightforward task. For the average demolition contractor, marketing involves contacting buyers and requesting their best prices for a given item. Terms such as price, quality, preparation requirements, shipping methods, and delivery schedules are typically part of any purchase agreement. It is sometimes necessary for the demolition contractor to advertise in publications likely to reach potential customers. The Internet is becoming a valuable tool for reaching a worldwide customer base. Most demolition companies that have been in business for a number of years will maintain a list of likely buyers for various salvageable items.

H. STUDY QUESTIONS

1. What type of debris is generally referred to as materials that have little or no net value as recycled materials?

2. What two major categories is scrap metal from demolition projects separated into?

3. When did serious recycling of concrete and CMU begin?

4. What types of petroleum products are pumped from their containers and reprocessed for use as fuel?

5. In the future of recycling, what will the regulations likely force the demolition contractor to do regarding recycling/disposal?

6. What is sustainable design and development? How does it impact the demolition contractor?

7. What programs are available to certify that a building's design, construction, and performance allow society to meet the needs of the present generation without compromising the ability of future generations to meet their needs?

CHAPTER 11

EXPLOSIVE DEMOLITION

A. INTRODUCTION

Almost every weekend the public is entertained by a high-rise building being collapsed or, to use the more popular term, "imploded" by the use of explosives. In fact, there is a misconception by most of the public that the word "demolition" means the use of explosives. While the use of explosives plays an important role in the demolition industry, most demolition is accomplished using heavy equipment.

For the most part, explosive demolition is performed by companies that specialize in implosions and other types of blasting work and most often work as subcontractors. The prime demolition contractor is a firm that has contracted to perform the entire project, which may include such tasks as removal of hazardous materials, mechanical wrecking, hauling, and backfilling.

A common phrase used by the public is "blowing up a building." This is a misconception insofar as the operating force for implosion is simply gravity. The explosives are used for the purpose of weakening the structural supports to the extent that they no longer can bear the weight of the structure, and it falls to the ground. In today's world, the structural support for buildings and other structures (bridges, stacks, tanks, industrial equipment, etc.) are primarily constructed of concrete, steel, and brick. The method for bringing down a structure with the use of explosives depends on the materials used in its construction and how it is built. The site location and local regulations also play an important part in the decision to use explosives.

In addition to bringing down high-rise buildings, explosives may be used to cut and fell steel structures such as bridges and industrial equipment. In some cases, explosives are used to break up heavy concrete foundations, fell stacks, and other specialized demolition tasks.

B. WHEN TO CONSIDER USING EXPLOSIVES

There are many reasons for using explosives, depending upon the circumstances of the project. Below are listed some of the more common reasons explosive demolition may be the ideal method:

- Because of location, condition, or other factors, explosive demolition may be the safest method for bringing down a building or other structure.
- The cost to fell a building or other structure by use of explosives may be less expensive than demolition by conventional means.
- A building or other structure is of sufficient height to make conventional demolition from wrecking staged on the ground too difficult and overly expensive.
- A very tight work schedule may provide incentive for the owner to either require or request that the demolition contractor use explosives.
- A building or other structure that has suffered prior damage due to fire, earthquake, hurricane, or other cause may not be stable enough to safely be demolished with heavy equipment. For example, the explosion damage caused by the bombing of the Alfred P. Murrah Building in Oklahoma City made that structure very unstable.
- A municipality may not want a local area to be subjected to weeks of conventional demolition and would encourage a one-time implosion event.

- Politics sometimes enter into the decision to use the implosion method. It makes a great show. Such occurrences have been popular public events and at times are used as fund-raisers.

While use of explosives may often be the fastest, most economical, and safest method for bringing down a structure, there are a number of factors that must be considered before making a decision to use explosives. Many specifications prohibit the use of explosives for a variety of reasons. Sometimes this is the result of lack of information as to whether or not explosives can be safely used.

Below are listed some of the more common reasons explosive demolition may not be used:

- Existing laws or regulations may prohibit bringing explosives onto certain sites for security reasons. A good example could be a military installation, nuclear power station, or petrochemical plant.
- The structure to be demolished with explosives is in a location such that the resulting debris pile from imploded structures might negatively impact adjacent structures, roadways, etc.
- The structure to be demolished with explosives is in close proximity to critical underground utilities, tunnels, subways, etc., and the resulting impact vibrations could damage them.
- Usually there is a large one-time cloud of dust created by the implosion of a structure. This may not be allowed in certain areas such as in close proximity to a hospital or school.
- Sometimes it is not financially wise to implode a building if mechanical demolition is less expensive and time is not a critical factor.

C. PREPARATIONS FOR USING EXPLOSIVES

As with all methods of demolition, proper preparations are perhaps the most important factors in assuring that the project will be successful. Each project destined to be demolished by the use of explosives is unique due to applicable regulations, location, type of construction, schedule, and methods required for the project to be a success.

In general, the proper implementation of the following preparations is instrumental to the success of an explosive demolition project:

1. A thorough review of federal, state, and local blasting regulations to ensure that the planned explosive demolition is permissible and the implosion demolition plan meets all applicable requirements.
2. Organize a meeting with the local public officials who will have the authority to grant an explosives demolition permit.
3. Prepare a detailed Explosives Demolition Work Plan to present to the owners, public officials, and other stakeholders. As a minimum, the plan should include the following:

- Identify with drawings which building and which structural components of the building will be imploded (cut or broken up with explosives).
- Describe which floors or levels will be imploded (a common practice is to shoot two or more floors/levels to increase breakage of materials and control the location of the resulting debris pile).
- Describe the means and methods that will be used for transporting, delivering, and temporarily storing the explosives on-site, as well as for disposing of excess explosives after the blast.
- Describe what building features need to be removed to provide access for blasting preparations.
- Plan and implement a test blast to determine the optimum amount of explosives to be used.
- Describe how protection will be used to contain pieces of construction materials from becoming dangerous projectiles referred to as "fly."

- Identify various walls and construction features such as stairwells, elevator shafts and rails, water standpipes, and any other building features that need to be weakened to reduce their ability to resist the felling of a structure in a planned direction.
- Identify where blast holes must be drilled in the case of concrete structures.

Figure C.01. Column drilled for explosives.

- For steel structures, identify where steel columns and bracing must be weakened as well as where and how the linear-shaped demolition charges are to be affixed to the structural steel supports.
- For steel bridges, identify where bridge members must be weakened in addition to where and how linear-shaped demolition charges are to be affixed to the members.
- Describe methods for protecting the public from access to any areas that may be impacted by either the preparations for implosion or the actual implosion operation.
- Describe how employees are to be protected while preparing the structure for explosive demolition.
- Describe the methods for photographing and conducting a visual survey of conditions before the implosion.
- Describe the type and location of vibration monitors and who will be responsible for their placement, readings, and producing a report of the results.
- Describe how still photos and videos will be used to show the "before and after" conditions.
- Describe how the site will be cleaned up after the blast.

D. IMPLOSION, THE SPECTACULAR USE OF EXPLOSIVES

Those able to witnesses an implosion firsthand are awestruck as the charges are initiated and the building comes crashing down in a huge cloud of dust. Often pre-demolition parties are held and money is raised for charities by auctioning off the rights to push the detonator switch.

Some of the important considerations that must be taken into account when planning the implosion of a building or other structure are among those listed below:

Controlling "Fly"—One of the major problems with imploding a structure in urban areas is the prevention of "fly" (pieces of debris that may be expelled by the blast) from injuring people or damaging property. During the preparation stages discussed previously, columns loaded with explosives are wrapped with materials such as geotech fabric and fencing that greatly reduce the possibility of fly.

Detonation of Charges—Critical to a successful implosion is the proper installation and timing of the detonations. Installation of explosive charges must be very organized and follow the carefully designed Explosives Demolition Work Plan that has been approved by the owners' representative. The timing of the detonations is critical to controlling the direction and manner of the collapse of a structure. This is achieved by the use of millisecond time delays in the firing circuits. These delays also reduce the force of shock waves, ground vibrations and noise.

Crowd Control—Most implosions are done shortly after daybreak on a Saturday or Sunday morning. This timing tends to reduce the size of the crowds and has the least impact on traffic. The areas around the perimeter of a building to be imploded are barricaded, and signs are placed warning people to stay away. It is common practice to hire off duty police to assist in keeping the spectators behind the barricades during the blast.

Dust Control and Cleanup—As a structure is collapsing, huge clouds of dust are created, and there is no effective way to keep the dust from coating buildings and pavements downwind of the shoot. It is common practice to cover air intakes and other openings of buildings expected to be in the path of the dust cloud. Water can be used to pre-wet the impact area and to reduce the time that the dust remains airborne. The contractor must be prepared to sweep streets, clean the dust from adjacent structures, and remove any temporary protection from nearby buildings.

Figure C.02. Twelve-story building ready to shoot.

Figure C.03. Loading test column with dynamite.

Figure C.04. Test column wrapped for shoot.

Figure C.05. Column after test shot.

E. EXPLOSIVE DEMOLITION OF STEEL STRUCTURES

The process for the demolition of steel supported structures is quite different from the processes used to explosively demolish concrete structures. Instead of placing charges in holes bored in the concrete columns, the columns of steel structures are prepared as follows:

- Shear walls, such as stairwell walls and elevator shafts, are demolished with machines to reduce the resistance of the structure to move in the desired manner.
- Standpipes, elevator shafts, and any other construction on the floors to be shot are cut to eliminate their ability to resist collapse.
- A column is weakened by cutting away portions of the column with torches.
- A linear shaped charge is a "V" shaped copper lined container loaded with explosives and a self-contained detonator. The enormous pressure generated by the detonation of the explosives projects the copper liner to form a continuous, knife-like jet of molten copper. The jet cuts any material in its path, to a depth dependent on the size and materials used in the charge. Shaped charges are usually used in tandem so that they remove a four to five foot length of the steel member, thus weakening the structure. A small separate charge may also be attached to the piece being cut to "kick" it out of the way after it is cut.
- The charges are timed to detonate in a precisely planned manner so that the structure collapses in the proper sequence, allowing the structure to fall in the desired direction.

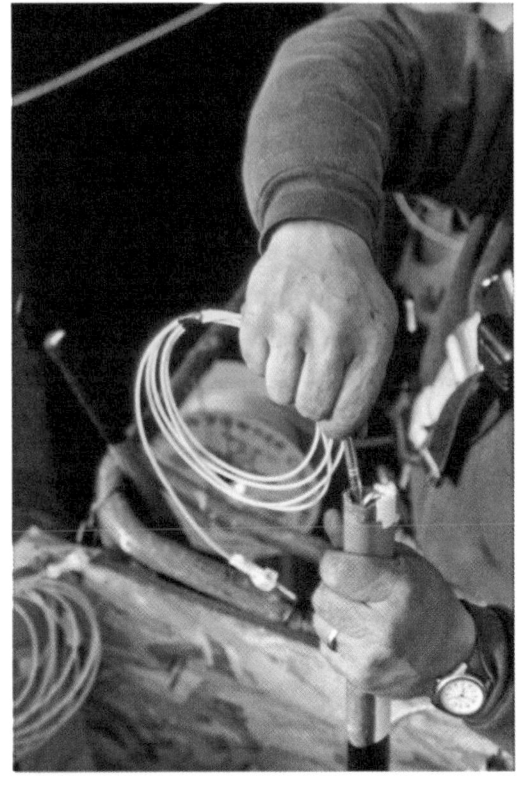

Figure D.01. Placing blasting cap in the charge.

Figure D.02. Loading the charge.

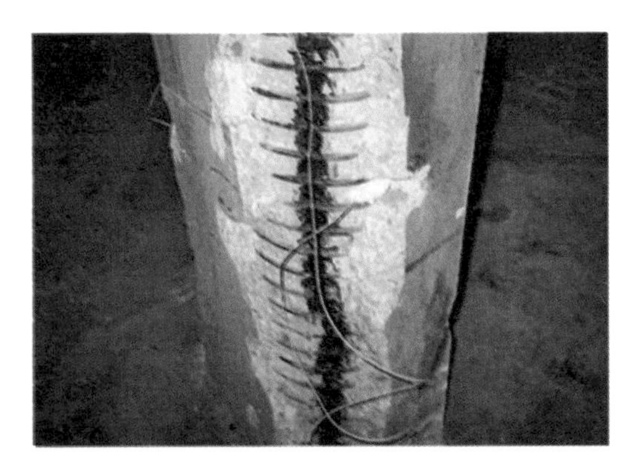

Figure D.03. Charges sealed with expanding foam.

Figure D.04. Cooling tower diagonal columns shot in two places. Detonation of explosives timed to control fall.

Figure D.05. Visible smoke shows cuts made in upper structure to control direction of fall and to break up the concrete.

Figure D.06. Complex shoot of 25,000 ton major sports stadium using 4,700 pounds of explosives in 5,900 holes. Detonated in six sections to avoid excess ground shock.

Figure D.07. Resulting debris pile only twenty-three feet high and the concrete was mostly separated from the reinforcing steel.

Figure D.08. Implosion of a downtown hotel.

Figure E.01. Shaped charge on cast steel.

Figure E.02. Shaped charge with dynamite "kicker."

Figure E.03. Steel building implosion.

Figure E.04. Las Vegas hotel implosion.

F. EXPLOSIVE DEMOLITION OF FOUNDATIONS

Explosives have been used for many years to break up building, machinery, and bridge foundations. The process is similar to that used in the mining industry. Holes are drilled in a planned pattern, and charges are primed and set into the holes. The rapid expansion of the explosive gases causes the concrete to shatter. For extensive foundations, large pneumatic drills are used to create the blast holes. Smaller foundations or those that are located in areas that are inaccessible to drill rigs are drilled with handheld jackhammers.

Where permissible, explosives are often used to break up underwater foundations. This can be done by drilling the blast holes from barge-mounted drilling rigs and loading the holes by using divers. Another technique is the placing of specially shaped charges right on top of the underwater foundation. The mass of water resists the explosive force and forces a significant percentage of the blast shock wave into the concrete, thus causing a fracture.

Figure F.01. Track-mounted rig drilling holes in foundation.

Figure F.02. Massive above ground machine foundation covered with soil and woven steel blasting mats to reduce "fly."

Figure F.03. Results of foundation blast shown in Figure F.02.

G. EXPLOSIVE DEMOLITION OF STACKS AND TOWERS

For many years large smoke stacks constructed of concrete or brick have been felled using explosives. Depending upon the planned direction of the fall, the stack base is drilled in a carefully planned pattern to a depth of about one-half of the thickness of the shell. Charges are loaded, and the holes are packed with sand, much like the process used to shoot building columns. Sometimes the stack is prepared by using hydraulic hammers to cut a wedge-shaped opening (referred to as a "bird's mouth") at the base of the stack with its center located in the direction of the fall. The blast holes are then drilled around the "bird's mouth" so that when the blast is initiated, enough material is removed to allow gravity to take over.

Figure G.01. Shooting a stack.

Figure G.02. Four stacks shot in sequence.

H. SPECIALIZED EXPLOSIVE USES

The demolition contractor may occasionally be required to perform unusual tasks that require the special use of explosives. A few examples of such work are listed below:

- Cut off a steel piling below the mud line
- Use small charges to break loose refractory bricks in kilns and furnaces
- Break up underwater concrete foundations and piers
- Drop a counterweight from a bascule bridge

Figure H.01. Bascule bridge 700 ton counterweight shot and dropped into bed of soil and used heavy equipment tires to reduce ground vibration.

I. STUDY QUESTIONS

1. Provide three examples of why using explosives for demolishing a structure is safer/smarter than general demolition by conventional means.

2. Name three key parts of a detailed Explosives Demolition Work Plan to be prepared prior to any demolition explosion and provided to the owners, public officials, and other stakeholders.

3. Regarding implosion and dust control, what is common practice to reduce and contain the time that the dust remains airborne?

4. Regarding explosion demolition of steel structures, what steps are taken instead of simply placing charges in holes of the concrete columns?

5. Name a few specialized explosive uses.

6. In the blasting process, why is the dynamite placed in a drilled hole and the hole then filled in with sand or expanding foam?

CHAPTER 12

DISASTER RESPONSE

A. INTRODUCTION

As we go about our normal activities, the forces of nature and acts of man sometimes interfere with our lives in the form of a disaster. Natural disasters, whether floods, hurricanes, tornados, earthquakes, structural collapse, tsunamis, volcanic eruptions, or other events, have been part of the human experience since records have been kept.

As the societies of mankind become more complex, man-made disasters have surfaced. War, fires, explosions, failed infrastructures such as levies and bridges, and a host of other calamities have occurred and will again.

In recent years, the ugly specter of "intentional disaster," commonly referred to as terrorism, has become a serious threat. The attacks by a small band of determined terrorists upon the United States on the morning of September 11, 2001 created a large loss of life, enormous property damage, and the disruption of many of the normal activities of modern society.

This chapter reviews the means and methods that the demolition industry provides to government agencies and private entities to help mitigate the damages created by natural or man-made disasters.

Figure A.01. Heavy equipment used in fire scene response.

B. THE ROLE OF THE DEMOLITION CONTRACTOR

Fire departments, assisted by other branches of government and utility companies are traditionally the *first responders* to either a natural or man-made disaster. These *first response teams* are critical to saving lives and keeping order.

As a result of the terrorist attack of 9/11, President George W. Bush's First Responder Initiative put forth \$3.5 billion for the needs of first responders. A new government Cabinet Post was established, Department of Homeland Security, which brought together several existing government agencies for the purpose of preventing and responding to all types of disasters. This legislation has enabled communities throughout the United States to train personnel and stockpile emergency equipment and material likely to be needed during first response efforts.

The importance of utilizing the services of the demolition industry to work with first responders was dramatically demonstrated by the terrorist attack of 9/11, the anthrax attack on several offices in the U.S. Hart Senate Office Building, and the domestic terror attack on the Alfred P. Murrah Building in Oklahoma City, Oklahoma. Also of significant importance has been the response of the demolition industry to natural disasters such as those of the 1989 and 1991 earthquakes in California and hurricanes Andrew and Katrina that ravaged the Southeast United States.

At the very onset of a disaster, the experienced demolition contractor can be called upon to quickly respond to the site with equipment and personnel to assist the fire, police, and emergency medical personnel as needed. Once the immediate dangers to life and property are brought under control at a disaster site, the demolition contractor is the best equipped and most experienced type of construction organization to continue the process of removing damaged structures and cleaning up the site. Listed below are the primary examples of why the demolition contractor is the best qualified to deal with disasters:

- Personnel employed by a demolition contractor are experienced in dealing with collapsed structures.
- The demolition contractor has equipment designed for selectively handling rubble from damaged structures.
- Nearly all major metropolitan areas of the United States and Canada are served by one or more demolition firms that have the experience and capabilities to undertake a role in disaster response.
- Most demolition companies have personnel trained in the removal and handling of hazardous materials, which often complicate rescue and removal activities.

Figure C.01. Training with a fire department.

C. TRAINING

Demolition contractors, by the very nature of their trade, work in conditions similar to those that could be expected at a disaster site. It is important that contractors wishing to play a role in disaster response be trained to work with other responders.

Many demolition contractors are experienced in working with traditional first responders and are able to provide employees trained in a variety of tasks typically needed at disaster sites. Such training may include the following:

- Working with first responders as part of a team
- Establishing chain of command
- Establishing exclusion zones
- Recognition of possible hazardous materials
- Removal of hazardous materials
- Entering confined spaces
- Fall protection
- Use of proper Personal Protection Equipment (PPE)
- Rigging
- Dust control

In order to have a dependable working relationship with local first responders, it is important that the demolition contractor has a previously established means of communication with the emergency response team. In addition, the demolition firm must be able to quickly communicate with its own key personnel so that mobilization to a site can be expedited.

Figure D.01. Emergency demolition of earthquake damage.

Figure D.02. Fire scene requiring emergency demolition.

D. COORDINATION WITH OTHER RESPONDERS

In most cases, a demolition firm will be called to the site of a disaster by a local fire department. Once a demolition contractor has been notified of a disaster, the coordinated effort to maximize the skills offered by the experienced firm may include any combination of the following tasks:

- Mobilization of personnel in a pre-planned system for communication
- Emergency mobilization of equipment to be coordinated with state and local transportation departments and coordinated with local police
- Provide cranes and/or excavators and loaders to remove debris for access of emergency medical personnel, firefighting equipment, and utility companies
- Provide rigging services for removal of structural components and other items
- Clear roadways
- Isolate burning areas to aid in firefighting
- Provide dust suppression
- Demolish structures in danger of collapse
- Provide high-reach access with manlifts and cranes
- Provide pumps for dewatering

Several of the nation's larger demolition firms participated in the demolition of the remains of the World Trade Center buildings and the removal of over 1,000,000 tons of debris.

E. STUDY QUESTIONS

1. Describe the function of a disaster "first responder." What governmental agencies provide disaster first response teams?

2. List primary examples of why demolition contractors are uniquely qualified to help mitigate the damages that result from natural or man-made disasters.

3. What contractor employee training is required before a demolition contractor can effectively assist in the response to a disaster?

4. Why is an established means of communication with first responders necessary before a demolition contactor can effectively participate in disaster response?

5. What disaster response tasks are demolition contractors typically qualified to participate in?

Figure D.03. Demolition contractors working to clear the site of the September 11, 2001 terrorist attack in New York City.

Figure D.04. Tornado cleanup.

CHAPTER 13

PROJECT MANAGEMENT

A. INTRODUCTION

Project management is an important part of every type of major construction activity. The demolition industry, however, has some unique requirements for project management that are not required in other types of construction work. Demolition contractors, in general, all have their own definition of what constitutes project management and of their individual requirements for the position of the project manager (PM) in their firm.

For purposes of this book, a typical format will be used for both the activities of project management and the duties of the PM. Project management may be briefly defined as the organization and implementation of those administrative tasks required to plan, start-up, maintain, and close-out a demolition project. The PM is the individual who has the responsibility to either self perform or supervise others in the administrative tasks required for project management.

B. PLANNING THE WORK

One of the key elements of a successful demolition project is development and implementation of a comprehensive management plan that establishes the framework for how the work will be controlled from start to finish. There are internal and external controls that must be accounted for within the plan. Formalizing these controls in turn allows the PM to manage the work to both company requirements and expectations (internal controls) and the client's requirements and expectations (external controls).

The management plan should encompass the administrative and technical requirements of the work, as well as codify the appropriate project controls and quality assurance measures that will be used to manage risk. The management plan should further include the mechanisms that will be used to communicate the requirements and controls to the project superintendent and the rest of the project team. This in turn provides the PM a means to evaluate and measure the team's ability to manage costs, schedules, safety, and the scope of work.

C. PROJECT START-UP

The PM must analyze each demolition project and determine what tasks will be required to move the company from a notice of successful bid to being able to start work. Listed below are tasks that are commonly performed by the PM or personnel working under his or her direction.

Contract Review—It is important to thoroughly read and understand the contract and be sure it accurately describes the work and the conditions set forth in the bidding documents. It may be necessary to have an attorney review confusing language and/or to evaluate if there are discrepancies or changes between the contract documents and the bid documents.

Subcontract Review—The PM usually has the responsibility to write subcontracts and verify that subcontractors have submitted required insurance documents and bonds as well as to secure all required signatures. It is imperative that the subcontract documents account for any flow-down provisions from the prime contract, or that any revisions to the flow-down requirements are documented and approved. The PM should manage the prompt delivery of any subcontract submittals that may be required.

Bonds and Insurance—The PM should carefully review these documents to make sure they comply with the contract specifications. It is important to note that if there is a request for "additional insureds" that the insurance carrier, corporate attorney, or management approves of such coverage.

Usually bonds and insurance are obtained in the proper form from the firms' providers and are delivered to the owner before work can start. In some instances, it may be of value to have the surety representative review the contract documents and scope of work prior to posting the bond.

Preliminary Lien Notices—In some states this document must be filed with the governing agency prior to the start of work or within a specific time frame. This filing will protect the contractor's lien rights if the customer and/or a lower-tier subcontractor do not pay as agreed in the contract(s). (Note: A lien is secured on real property by a contractor who has performed work on that property in order to ensure payment for labor and materials. A lien is a legal document that encumbers any transfer of real property until the contractor is paid for the work.)

Contract Submittals—Depending on the contract specifications, there may be several submittals required before the demolition contractor can begin work. The PM is responsible for managing the production of submittals in a timely manner. Some submittals can require long lead times to ensure that they are approved before the work begins. Examples of typical submittals are listed below:

- Schedule—Cost-loaded schedules are typically required (a cost-loaded schedule shows the cost of an item as well as the time frame and duration allowed)
- Site Specific Health and Safety Plan
- Site Specific Environmental Plan
- Demolition Work Plan
- Material Handling/Disposal/Recycling Plan
- Dust Control Plan
- Temporary Facilities and Controls Plan
- Engineering Survey—OSHA 1926.850(a)
- Storm Water Pollution Prevention Plan (SWPPP)
- Quality Control Plan—(Note: not often required for a typical demolition project)
- Project photographs
- As-built drawings—(primarily needed for showing locations of utility caps and remaining foundations)

Permits/Notifications—Each project may be subject to various permits and/or notifications. Depending on the location of the project, the work may be subject to federal, state, or local permit and notification requirements. Examples include:

- Underground Service Alert
- Demolition Permit(s) (also known as Building Permits in some jurisdictions)
- Asbestos Removal Notifications (NESHAP Notification)
- Other Hazardous Materials Removal Notifications
- Excavation Permits
- Underground Tank Removals
- Sewer Caps
- Street and Sidewalk Closure Permits
- Extended Work Hours Permits, etc.

All permits must be obtained in a timely manner to avoid delays and penalties. Additionally, if the project is located in a jurisdiction where an outside entity must perform utility isolations and terminations (e.g., gas or electrical services), the notifications or service requests must be submitted in advance to ensure timely completion of the service request. Permits are often tied to a sign-off that the work is complete from such entities. Failure to complete the service request or to account for its impact on the overall project schedule could result in significant delays.

Meetings—Meetings are an important component for the success of any project. Without proper design and control, meetings can be a significant drain on time and resources. The PM should always be prepared for meetings by establishing and sticking to a concise agenda structured to the needs of the project. The agenda should at a minimum allow for review of any carryover action items from previous meetings, a review of the work progress, observations, problems and decisions, the identification of problems that impede progress, review and status of submittals, health and safety issues, corrective actions, maintenance of progress schedules, as well as changes and their effect on schedule or coordination.

Depending upon the project there may be two or more pre-work meetings required during the project start-up period. Some clients will expect that weekly, bi-weekly, or monthly meetings follow throughout the course of the project. The PM will usually organize the company participation in such meetings. Again, care must be taken to define the objectives of the meeting and to stay on track with the pertinent topics. If required, meetings should be structured in a flexible manner to allow for attendees to participate only on an as-needed basis. For example, if the demolition work is part of a large redevelopment project, it may not be necessary for the demolition PM or superintendent to attend the entire meeting. It is still recommended that the demolition PM devise and use a demolition-specific agenda to communicate the work progress. Other tips for pre-work meetings include:

- Because the project superintendent is instrumental in completing the Engineering Survey, they are valuable in identifying equipment and resource needs, arranging for an on-site office, if required, etc. It is important that the project superintendent be brought into the process as early as possible to assist the PM with such tasks as understanding and completing prerequisite tasks required to ready the site for demolition (e.g., utility isolations or re-routes, areas to be protected, work by others). It is desirable to have a meeting early in the schedule with the owner's representative(s) to review the scope of work, schedule, and interfaces of work with other contractors, etc. At this point any misunderstandings or changes to the work can be addressed and dealt with before work begins.

- It is common for the demolition contractor's PM to schedule a pre-work meeting with any government or regulatory agencies that may be involved in activities such as permitting and inspection. Clients often wish to participate in this meeting, particularly if there are subsurface contamination concerns associated with the work, or if the demolition work is part of a larger development project.

- Meetings with subcontractors should be scheduled by the PM and attended by both the PM and superintendent.

- Depending upon the location, labor contracts with unions may require the contractor's PM to arrange a pre-job meeting with local labor unions. If certified payrolls are required as part of the contract or a union agreement, arrangements must be made to fulfill such requirements.

Review Any Special Employment/Subcontractor Requirements—Some contracts will have Minority Business Enterprise (MBE), Women's Business Enterprise (WBE), Disadvantaged Business Enterprise (DBE), or other subcontracting and employment hiring goals. The PM should be responsible to ensure that contractual requirements for these categories be fulfilled. The PM should verify documentation of good faith efforts to fulfill such requirements.

Project Mobilization—The PM and the superintendent generally work together in organizing mobilization to the project. In addition to equipment transportation, project mobilization may require setting up temporary facilities and controls (e.g., temporary offices, lunch areas, toilets, provisions for fueling, etc).

D. PROJECT ADMINISTRATION

Liaison with the Owner's Representative—the PM is usually the primary contact between the demolition contractor and the owner's representative. Regular meetings may be held so that the PM can keep the owner advised of such activities as schedule changes, progress payments, safety concerns, and personnel changes. Project meetings are critical to the project communications process and must be accurately documented to ensure that information and action items are available to the owner's representative and to non-attending team members. Meeting minutes should be issued within twenty-four hours of the meetings. All communications of any kind with the owner's representative should be documented in writing and be a part of the project records.

Liaison with the Project Superintendent—A well run, efficiently managed project is the result of good cooperation between the PM and the project superintendent. The PM should verify that the superintendent has been provided with all the proper notifications, permits, and any other documentation that needs to be posted on the jobsite. Both individuals need to be aware of the project needs on a daily basis, and the PM should be ready, whenever necessary, to provide assistance to the superintendent. Regular meetings between the PM and the project superintendent should be treated with the same diligence as any meeting between the PM and the owner's representative. An agenda should be established, and all meeting decisions should be documented and made available to appropriate team members.

Daily, Weekly, and Monthly Progress Reports—Depending on the size of the project and the company's policy, the PM usually manages the production of periodic reports on the progress of the work and the profit/loss projections. Included are safety reports, other incident reports, and cost reports. It is important to establish at the beginning of the project whether an owner expects any type of closeout document or report, as well as as-built drawings. These requirements may not be formalized or communicated until late in the project. It can be extremely difficult to reconstruct activities that took place early on in the schedule, especially with respect to as-built drawings. Determine the expectations early to manage and maintain the information that will support the development and submittal of a comprehensive closeout document. Moreover, if there are expectations that require additional compensation, negotiate this change early on.

Invoicing—The PM should plan ahead each month to be sure that he or she initiates the proper documentation for the invoicing process. It is recommended that early in the project, preferably during a pre-work meeting, an example invoice be submitted for approval. Establish any requirements prior to submission of the first invoice to avoid invoice rejection, which in turn delays payment. If retention applies to the invoice, be sure that the invoice clearly identifies that the retention is accounted for. Conversely, if retention does not apply, clearly state that on the invoice to avoid invoice rejection. The PM should establish a schedule for preparation and review of each invoice. Late invoicing is indicative of poor administrative control of a project, and can result in late payments. This can cause unnecessary problems as well as impact the profitability of the project.

Change Orders—The PM usually manages the request and information needed for submitting change orders. The company's estimator(s) are often involved in this process, and it is important that the PM obtains all the documentation needed for a change order in a timely manner. Contracts often have very specific requirements for the submittal and timing of change orders. The PM and/or the project superintendent must be diligent to comply with these requirements to ensure payment for change order work. In some instances, the change order is prefaced with a field change directive or field order, which allows for work to be started prior to issue of a formal change order. This is common when an unexpected condition is encountered and must be handled immediately. A field change directive and/or field order must still be documented to establish why the work is needed, who is requesting the work, how the work was authorized, and how the demolition contractor will be compensated for the work.

Labor Management—For contractors working in their normal geographical area, labor management is a rather simple process. Management, the PM, and project superintendent may all be involved in the assignment of labor for a project. For out-of-town work, the union contractor may have to negotiate with labor unions to bring in personnel from other areas and for hiring union members in the project area. The PM and superintendent will often work together to arrange for local labor in an unfamiliar location.

The project superintendent is usually responsible for the daily management of personnel and will hire and lay off workers as the project requires.

Subcontractor Management—The PM is usually responsible for overall management of any subcontractor activities, whereas the superintendent provides daily oversight and coordination. The PM's responsibilities may involve schedule changes, employee problems, subcontractor invoice approvals, among other issues. Any changes to the subcontract agreement, scope of work, or schedule must be documented in writing and approved by the PM.

Salvage Sales—Whereas routine salvage of metals and small ticket items are usually handled by the project superintendent, the PM should be involved in major sales. It is very important that the PM manage sales that will have a major impact on the profitability of a project. This may involve negotiations and contracting with buyer(s), arranging transportation, managing weights/volumes, quality control duties, establishing payment terms, and assuring that payment terms are met.

Equipment Utilization—The selection and utilization of major equipment for a project is typically decided by the contractor management. Equipment may be either contractor owned or leased. The PM and the superintendent are also often involved in this process and may typically be responsible for the acquisition and utilization of minor equipment.

Material Purchases and Rentals—All major material and equipment rentals required for a project should be coordinated by the PM with the company's purchasing department or equivalent. Provisions for minor expenditures should be made to support the project superintendent to avoid delays (e.g., payments for fuel, gases, and tools).

Jobsite Inspections—On some projects, a full time PM is required. For other projects, the PM can manage without full time participation. However, it should be recognized that a PM cannot adequately perform his or her responsibilities without frequent visits to the jobsite. Whenever possible, these visits should be coordinated with the project superintendent on established meeting intervals. Such meetings should be formalized by agenda and documented as noted previously. Except in an emergency or by prior agreement, it is usually not the duty of the PM to change any jobsite procedures, equipment, and personnel assignments—that shall be the responsibility of the project superintendent. Such responsibilities should be established in the management plan previously discussed.

Jobsite inspections can also include inspections by the owner's representative and/or by outside entities, including regulatory agencies. Such inspections can be scheduled inspections or they can be surprise inspections. All such inspections should include participation of the PM and/or project superintendent, and the inspection should be documented. During the jobsite inspection, common sense and a little diplomacy will go a long way toward achieving a safe, efficient, and properly managed project.

Visual Jobsite Records—The PM usually establishes the means and methods in the management plan for maintaining a jobsite record. This includes the requirements for taking videos and photographs of the work progress and/or any conditions or circumstances that may impact the work. Any safety-related issues, contamination, or other incidents should be photographed and documented in the project records. Some contracts have requirements for photos and videos, and the expectations for these should be established prior to the start of work (e.g., time-lapse photography).

Jobsite Production Statistics—It is recommended that the PM manage the collection of production data for use by the company on future cost estimates. This task can usually be performed by office personnel with oversight by the superintendent or PM.

Record Maintenance—The PM shall ensure that all job files are kept up-to-date and that all critical reports such as safety records, daily reports, payroll, physical exams, and testing are properly maintained and stored. Certain medical records and/or safety-related records have a much longer retention requirement than other documents, and the PM should be certain that they are properly stored.

E. CONTRACT CLOSE-OUT

As the project nears completion, there may be a host of contractual obligations to prepare for closing out the contract. Typically, the PM and/or the project superintendent will have the following tasks to complete:

Develop Punch List—Once the work is complete, the PM and project superintendent will conduct a walk-through of the site with the owner's representative. A "punch list" will be prepared identifying any outstanding elements that require attention prior to the work being accepted as final.

Prepare Certificate of ***Substantial*** *Completion*—Once the demolition work is substantially complete, the PM will prepare a "Certificate of Substantial Completion" for signature by the owner's representative. This certificate should provide for final acceptance of the work pending addressing punch list items.

Prepare Certificate of ***Final*** *Completion*—Once the work and punch list items are fully complete, it is a good practice to prepare a "Certificate of Final Completion" for signature by the owner's representative. This certificate will provide for final acceptance of the work by the owner.

Submit Letter of Certification—If required by contract, or if the project was subject to preliminary notices, once the Certificate of Substantial Completion has been signed, the PM may prepare and submit a "Letter of Certification" or "Letter of Construction Completion" to all interested parties, the owner, and to subcontractors.

Demobilize—Upon completion of punch list corrective actions, any remaining equipment, materials, or personnel will be demobilized from the project site.

Contract Billings, Including Change Orders—The PM is usually responsible for ensuring that all work has been billed, including retainage.

Salvage Sales—The PM should verify that all sales payments have been made and that unsold items are transported to inventory storage or disposed of.

Submittal Closeouts—The PM should verify that all submittal closeouts have been completed, including as-built drawings as required by the specifications.

Bonding Company—The PM should notify the bonding company when the project has been completed and accepted by the client.

Special Notifications—The PM usually is responsible for notices of completion of asbestos abatement, completion of storm water protection measures, completion of removal of underground and above ground storage tanks, among other items.

F. SUMMARY

Every demolition contractor will have developed his or her own programs and systems for project management and administration. This chapter was written with the understanding that the approaches to project management will vary significantly throughout the demolition industry. However, most companies will include the majority of the information discussed herein. All construction projects, regardless of how small or how complex, will benefit from the implementation of appropriate planning and controls. In addition to helping the current project run smoothly, project controls will help to establish the mechanisms by which future project improvements can be made.

The duties of a project manager will differ greatly from company to company and will depend on the requirements of individual projects. Some smaller demolition contractors operate without the services of a designated PM. The PM duties of these companies may be handled by the company owner or office personnel.

G. STUDY QUESTIONS

1. During project start-up what is the first thing that needs to be reviewed?

2. What key topics do the project meetings need to cover?

3. Who on the project is typically responsible for the overall management of the subcontractors?

4. What types of documents/records is the PM responsible for keeping up-to-date?

5. Explain the purpose of a "Punch List" on demolition project sites.

GLOSSARY OF TERMS, ABBREVIATIONS, AND ACRONYMS

ITEM	DEFINITION
Abatement	Defined as “an interruption in the intensity or amount of something.” When environmental policy or regulation requires abatement activities, the governing policies typically define the term abatement and the corresponding activities that relate to the particular hazard for which the regulation was written.
ACM	Asbestos-Containing Material. Asbestos-containing building materials (ACM) are one of the most commonly recognized materials requiring removal prior to demolition. ACMs are found in construction debris in furnace and pipe insulation, insulation, mastic, floor tile, ceiling tile, siding, caulking, transite board, roof shingles, etc.
ANSI	American National Standards Institute
Asbestos	Mineral fiber that can pollute air or water and cause cancer or asbestosis when inhaled. EPA has banned or severely restricted its use in manufacturing and construction. Materials that contain asbestos are collectively referred to as ACM.
AST	Above-Ground Storage Tank
Backfill	Refers to the practice of returning soil to a demolition site during site restoration to preserve a specified grade and leave the site in proper condition for future use.
C&D Waste (Debris)	Construction and Demolition Debris. Materials generated as a result of construction, renovation, demolition, and/or removal projects.
CERCLA	Comprehensive Environmental Response, Compensation, and Liability Act was enacted by Congress to create a tax on the chemical and petroleum industries and provide broad Federal authority to respond directly to releases or threatened releases of hazardous substances that may endanger public health or the environment.
CFR	Code of Federal Regulations.

ITEM	DEFINITION
CMU	Concrete Masonry Unit. Commonly referred to as cinder blocks or concrete blocks. CMU construction is typically grouted at hollow CMU foundation stem walls and exterior load bearing walls where some hollow cores are filled with concrete reinforced with steel reinforcement bars.
Crushing	Refers to a process where masonry, asphalt, concrete, and/or stone material is fed through an impact or jaw crushing machine that crushes the material into much smaller pieces, which can then be reused as aggregate or backfill material.
Davis-Bacon Act	The congressional act that established the federal guidelines for wages to be paid to various classes of laborers and mechanics employed under federal contract. For more information about the Davis-Bacon Act, visit the Department of Labor Web site.
Deconstruction	Planned and controlled disassembly of a building that preserves the integrity of the building materials and components so that they can be reused or recycled. When the type of construction does not lend itself to "disassembly," the term deconstruction means the breaking apart of building elements into their more basic constituents (steel, crushed concrete, etc.) and processing for potential reuse and or recycling. Also known as "sustainable infrastructure removal."
Demolition	An engineering project to reduce a building, structure, paved surface, or utility infrastructure through manual and/or mechanized means, with or without the assistance of explosive materials to piles of mixed rubble or debris. Demolition usually provides the quickest method of removing a facility and segregates the debris or rubble into various components for recycling wherever practicable.
Diversion	The redirection of waste ordinarily disposed of in a landfill or burned in an incinerator, to a recycling facility, to a composting yard, or to another destination for reclamation or reuse.
Drop Box	A metal debris container delivered by truck on a tilting frame and "dropped" at the site by sliding or rolling the container onto a flat surface. Also referred to as a "roll-off" container.
Engineering Survey	Prior to demolition, OSHA Standard 1926.850(a) requires that an engineering survey of the structure must be conducted by a competent person. The purpose of this survey is to determine the condition of the framing, floors, and walls so that measures can be taken, if necessary, to prevent the premature collapse of any portion of the structure.
EPA	Environmental Protection Agency. For more information, visit the EPA Web site.

ITEM	DEFINITION
FAR	Federal Acquisitions Regulations.
Final Completion	Completion of all contractual requirements of a project.
First Responders	Those individuals in the early stages of an incident, who are responsible for the protection and preservation of life, property, evidence, and the environment.
Fixed Costs	Company costs that within a relevant range of project volume are not variable. These costs are considered to be the cost of doing business and must be paid regardless of the number or dollar value of project work undertaken. Also referred to as "overhead costs."
Fly	Pieces of debris that may be expelled by a blast of explosives.
Flow-Down Contract Provisions	Contract provisions in the agreement between the prime contractor and the subcontractor that require the subcontractor to abide by contract requirements contained in the contract between the prime contractor and the owner.
G & A	General and Administrative Costs
Geotech Fabric	Geotechnical fabric is a woven product typically used for soil, embankment, and roadway stabilization and separation. Also used to restrict "fly" in explosive demolition.
Grading	Manipulation of a section of terrain to match existing topographic features, slope for adequate drainage, or other landscape requirements.
Grinding	A technique used to support diversion where building materials are ground for alternative use. Common items typically include wood and gypsum that are ground to produce beneficial mulch, soil stabilization materials, soil amendments, etc.
Hazardous Waste	A waste material that generally meets RCRA requirements and conditions as a listed (predefined categories) waste or characteristic (exhibits certain characteristics) waste. Characteristic wastes must exhibit characteristics associated with corrosivity, reactivity, ignitability, or toxicity. For a more detailed definition, visit EPA Hazardous Waste Information Web site.
HAZMAT	Hazardous Materials—any solid, liquid, or gas that can harm people, other living organisms, property, or the environment. A hazardous material may be radioactive, flammable, explosive, toxic, corrosive, biohazardous, an oxidizer, an asphyxiant, an allergen, or may have other characteristics that make it hazardous in specific circumstances.

ITEM	DEFINITION
HEPA	High efficiency particulate air filter
"Hi-Reach"	Long-reach demolition machines are a development of the excavator with an especially long boom arm that is used exclusively for demolition. The high-reach demolition machine is designed to reach the upper stories of buildings in excess of one hundred feet in height to demolish the structure in a controlled fashion.
Indemnification Clause	Provision in a contract in which one party agrees to be financially responsible for specified types of damages, claims, or losses.
LBP	Lead-based Paint. Lead-based paint is not the only source of lead exposure in demolition, but due to its prevalence in much of the paint used until the late 1970s, it is a prominent lead hazard that requires proper environmental management.
LEED	Leadership in Energy and Environmental Design
Lien	A lien is secured on real property by a contractor who has performed work on that property in order to ensure payment for labor and materials. A lien is a legal document that encumbers any transfer of real property until the contractor is paid for the work.
Lump Sum Contract	A contract that specifies a fixed price for the completion of the project described by the contract terms. Also referred to as a "fixed price contract."
MCL	Maximum Contamination Level
MBE	Minority Business Enterprise (MBE), Women Owned Business Enterprise (WBE), and Disadvantaged Business Enterprise (DBE) are owner-enforced subcontracting and employment hiring goals.
NESHAP	National Emissions Standards for Hazardous Air Pollutants. For more information about NESHAP, visit NESHAP FAQs on the EPA Web site.
NPDES	The National Pollution Discharge Elimination System, administered by EPA, requires projects over one acre obtain a permit and devise a system to prevent pollutants conveyed by stormwater or snow melt from leaving the site.
NPL	EPA's National Priorities List of Superfund sites.
OSHA	Occupational Safety and Health Administration. For more information, visit the OSHA Web site.

ITEM	DEFINITION
Overhead Costs	Company costs that within a relevant range of project volume are not variable. Theses costs are considered to be the cost of doing business and must be paid regardless of the number or dollar value of project work undertaken. Also referred to as "fixed costs."
PEL	Permissible Exposure Limits establish maximum level of safe worksite exposure to hazardous materials.
POL	Petroleum, Oils, and Lubricants
PPE	Personal Protective Equipment
Prime Contractor	Any contractor who has directly contracted with the owner.
RCRA	Resource Conservation and Recovery Act, designed to protect human health and environment by establishing a comprehensive regulatory framework for investigating and addressing past, present, and in some cases future environmental contamination at hazardous waste treatment, storage, and disposal facilities.
Recyclable Material	Materials that can be remanufactured into new products.
Recycling Facility	A facility that specializes in collecting, handling, processing, distributing, or reclaiming usable materials from a waste stream for reuse by others or remanufacturing into new products.
Retainage	Portion of the payment due to a contractor that is withheld until final inspection and acceptance of the work.
Roll-Off	A metal debris container delivered by truck on a tilting frame and "dropped" at the site by sliding or rolling the container onto a flat surface. Also referred to as a "drop box."
ROM	Rough Order of Magnitude
Selective Demolition	A careful demolition procedure whereby parts of a structure are removed while the primary structure is protected and remains intact.
SSHASP	Site Specific Health and Safety Plan. Development of an SSHASP is generally required for demolition and construction activities. For more detailed requirements of OSHA SSHASP requirements, refer to 29CFR1910.
SWPPP	Storm Water Pollution Prevention Plan, required on projects over one acre, describing the means and methods for controlling runoff from rainwater and snow melt.

ITEM	DEFINITION
Standards	Benchmarks promulgated by a regulatory agency, created to enforce the provisions of legislation.
Subcontractor	A contractor that contracts to complete work that is a partial fulfillment of the contact requirements between the prime contractor and the owner.
Submittals	Data, plans, reports and miscellaneous information to be submitted to the owner or prime contractor. Submittals are required by contract, primarily to verify that the correct procedures, processes, or products will be used on the project.
Substantial Completion	Completion of contractual requirements of a project to the point that the project can be used for its intended purpose. At substantial completion punch-list items may remain. Substantial completion may justify the payment of all but "retainage," especially in cases where tasks cannot be completed due to circumstances that are beyond the control of the demolition contractor.
Surfactant	Wetting agents that lower the surface tension of a liquid, allowing easier spreading. Used to make ACM easier to wet during removal.
T&M	Time and Materials. This is a common method of contracting.
TCLP	Toxicity Characteristic Leaching Procedure is an analysis relating to the potential public and environmental health risks posed by the liquids that percolate through landfills. The TCLP simulates conditions in the landfill and the reaction of these liquids as they come in contact with the landfilled solid wastes to determine the presence and concentration of contaminants identified by the EPA in the leachate.
"Third Member"	An attachment or additional element that works as an extension of the excavator boom and stick as opposed to replacing the stick. Attachments mounted directly to the boom are generally referred to as "second member" attachments.
Tipping Fee	Cost of disposal at a landfill. This is also referred to as disposal cost and is typically measured in $/tons or $/cubic yard.
USACE	United States Army Corps of Engineers
UST	Underground Storage Tank. A tank and any underground piping connected to the tank that has at least 10 percent of its combined volume underground.

INDEX